COMPREHENSIVE CHEMICAL KINETICS

COMPREHENSIVE

Section 1.	THE PRACTICE AND THEORY OF KINETICS (3 volumes)
Section 2.	HOMOGENEOUS DECOMPOSITION AND ISOMERISATION REACTIONS (2 volumes)
Section 3.	INORGANIC REACTIONS (2 volumes)
Section 4.	ORGANIC REACTIONS (5 volumes)
Section 5.	POLYMERISATION REACTIONS (3 volumes)
Section 6.	OXIDATION AND COMBUSTION REACTIONS (2 volumes)
Section 7.	SELECTED ELEMENTARY REACTIONS (1 volume)
Section 8.	HETEROGENEOUS REACTIONS (4 volumes)
Section 9.	KINETICS AND CHEMICAL TECHNOLOGY (1 volume)
Section 10.	MODERN METHODS, THEORY AND DATA

CHEMICAL KINETICS

EDITED BY

N.J.B. GREEN
King's College London
London, England

VOLUME 39

UNIMOLECULAR KINETICS

PART 1. THE REACTION STEP

N.J.B. GREEN
Department of Chemistry
King's College London
Strand, London WC2R 2LS
UK

2003

ELSEVIER
AMSTERDAM - BOSTON - HEIDELBERG - LONDON - NEW YORK - OXFORD
PARIS - SAN DIEGO - SAN FRANCISCO - SINGAPORE - SIDNEY - TOKYO

ELSEVIER B.V.
Sara Burgerhartstraat 25
P.O. Box 211, 1000 AE Amsterdam
The Netherlands

ELSEVIER Inc.
525 B Street, Suite 1900
San Diego, CA 92101-4495
USA

ELSEVIER Ltd
The Boulevard, Langford Lane
Kidlington, Oxford OX5 1GB
UK

ELSEVIER Ltd
84 Theobalds Road
London WC1X 8RR
UK

1st. edition 2003

Library of Congress Cataloging in Publication Data
A catalog record is available from the Library of Congress.

British Library Cataloguing in Publication Data
A catalogue record is available from the British Library.

ISBN: 0 444 41631 5 (Series)
0 444 50893 7 (Vol 39)
ISSN: 0069-8040

Transferred to Digital Printing 2009

COMPREHENSIVE CHEMICAL KINETICS

Volumes in the Series

Section 8. HETEROGENEOUS REACTIONS (4 volumes)

Volume 19	Simple Processes at the Gas-Solid Interface
Volume 20	Complex Catalytic Processes
Volume 21	Reactions of Solids with Gases
Volume 22	Reactions in the Solid State

Section 9. KINETICS AND CHEMICAL TECHNOLOGY (1 volume)

Volume 23	Kinetics and Chemical Technology

Section 10. MODERN METHODS, THEORY, AND DATA

Volume 24	Modern Methods in Kinetics
Volume 25	Diffusion-limited Reactions
Volume 26	Electrode Kinetics: Principles and Methodology
Volume 27	Electrode Kinetics: Reactions
Volume 28	Reactions at the Liquid-Solid Interface
Volume 29	New Techniques for the Study of Electrodes and their Reactions
Volume 30	Electron Tunneling in Chemistry. Chemical Reactions over Large Distances
Volume 31	Mechanism and Kinetics of Addition Polymerizations
Volume 32	Kinetic Models of Catalytic Reactions
Volume 33	Catastrophe Theory
Volume 34	Modern Aspects of Diffusion-Controlled Reactions
Volume 35	Low-temperature Combustion and Autoignition
Volume 36	Photokinetics: Theoretical Fundamentals and Applications
Volume 37	Applications of Kinetic Modelling
Volume 38	Kinetics of Homogeneous Multistep Reactions
Volume 39	Unimolecular Kinetics Part 1. The Reaction Step

Preface

Unimolecular reactions are in principle the simplest chemical reactions because they only involve one molecule. The basic mechanism, in which the competition between the chemical reaction step and a collisional deactivation leads to a pressure-dependent rate coefficient, has been understood for a long time. However, this is a rapidly developing field, and many new and important discoveries have been made in the past decade.

This project has grown during its realization, and it has been necessary to divide it into two volumes. This first volume is focused on the reaction step, and the second volume deals with the detailed description of the energy transfer step and its consequences, for example the fall-off of the rate coefficient.

In this volume, the first chapter is an introduction to the whole project, aiming to cover the introductory material necessary to understand the content of the detailed chapters, as well as a history of the development of the area, up to the development of statistical theories of the reaction step and the master equation description of the collisional energy transfer. Chapter 2 is a review of the modern view of the statistical theories, as embodied in the various forms of RRKM theory. This chapter deals with the problems associated with the application of the theory, both to reactions with a well-defined energy barrier, and to reactions without a barrier. Chapter 3 deals with the fully quantum mechanical view of reactive states as resonances. Because this topic is not covered in many books on unimolecular reactions, the chapter has a didactic introduction, before describing the various methods available for finding and characterising resonances, and reviewing the applications of the methods. This quantum picture is characterised by large fluctuations in the rate coefficient for states with quite similar energies, in contrast with the stepwise increase of statistical models. These fluctuations may be close to the statistical rate coefficient, but in some cases they are not, and the chapter deals with the reasons for this.

Volume 2 consists of two sections. The first section concentrates on collisional energy transfer – its experimental investigation, theoretical descriptions both from the point of view of classical trajectories and quantum scattering theory, and the simple parametric models that are often employed in applications to kinetics. The second section concentrates on the Master equation, a stochastic model which characterises the energy transfer collision by collision. The section also deals with the numerical and approximate methods for solving the master equation, as well as Monte Carlo methods for simulating the underlying stochastic process. An important strand running through the section is the application of the master equation to complex systems of coupled unimolecular reactions.

I would like to thank all the authors for the hard work they have put into the preparation of their chapters, and their patience during the long period it has taken to realise this project.

Nicholas Green

Contents

Chapter 1

Introduction

Nicholas J.B. Green [a]

[a]Department of Chemistry,
King's College London, Strand, London WC2R 2LS

1. INTRODUCTION

Unimolecular reactions are in principle the simplest types of chemical reaction that can occur in the gas phase because the reaction formally involves only one molecule. There are several different kinds of unimolecular reaction, but most can be considered either to be dissociation reactions, such as the dissociation of ethane to methyl radicals,

$$C_2H_6 \longrightarrow 2CH_3$$

or isomerization reactions, such as the *cis-trans* isomerization of but-2-ene. In order to undergo such a reaction the molecule must achieve a critical minimum energy threshold. In a thermal unimolecular reaction the necessary energy is supplied by collisions, usually with atoms or molecules of an inert 'bath gas'. However, it is also common for molecules to be formed in an energized state by chemical reactions, for example

$$^3CH_2 + C_2H_4 \longrightarrow C_3H_6^*$$

In this reaction, triplet methylene inserts in the ethene molecule to produce an energized cyclopropane molecule, which may then either isomerize to propene or may be collisionally deactivated to give cyclopropane. Molecules may also be activated photochemically. Although most photochemical processes involve more than one electronic state of the molecule, it is possible in some cases to produce molecules in their electronic ground states with high vibrational excitation, and these may subsequently isomerize or dissociate. A typical example is the photoexcitation of cycloheptatriene followed by its isomerization to toluene.

It has been known since the start of the 20th century that thermal unimolecular reactions obey first-order kinetics, and for some time this observation posed a problem. If molecules require energization to react, and collisional energization is a second-order process, how can the kinetics be first-order?

The first attempt to resolve this problem was the radiation hypothesis of Perrin [1], in which the molecule is assumed to be energized by absorbing thermal radiation (in the infra-red) emitted by the walls of the vessel. However it was rapidly realized that the intensity of thermal radiation was quite insufficient to explain the observed rates of reaction [2], although interestingly, under conditions of extremely low collision frequency in interstellar space, Perrin's radiation mechanism is now believed to be significant [3].

It has been generally recognized since about 1920 that thermal unimolecular reactions are activated by intermolecular collisions. The key idea is that the chemical reaction does not follow immediately after energization, but competes with collisional deactivation of the energized molecule. This idea, and the associated mechanism, were suggested independently by Lindemann [4] and Christiansen [5]. In the first step of this mechanism a molecule A is energized to A* by collision with a bath gas molecule M.

$$\mathrm{A} + \mathrm{M} \xrightarrow{k_1} \mathrm{A}^* + \mathrm{M}$$

Lindemann assumed that the rate coefficient for this process could be calculated from simple collision theory,

$$k_1 = Z_{\mathrm{AM}} \exp(-E_0/kT) \tag{1}$$

in which Z_{AM} is the second order rate constant describing the rate of $\mathrm{A} + \mathrm{M}$ collisions and $\exp(-E_0/kT)$ is the probability that the relative kinetic energy along the line of centres is greater than E_0, the threshold energy for the reaction.

The energized molecule may now either be deactivated by a further collision,

$$\mathrm{A}^* + \mathrm{M} \xrightarrow{k_{-1}} \mathrm{A} + \mathrm{M}$$

or react to form the product:

$$\mathrm{A}^* \xrightarrow{k_2} \text{Product}$$

Lindemann assumed that every collision of an energized A* molecule would be deactivating, so that the rate coefficient k_{-1} may be identified with the second-order collision rate constant Z_{AM}. The rate constant for the reaction step k_2 was to be determined from experimental results.

This kinetic scheme may be solved exactly by elementary methods. The reaction scheme is expressed as a set of two coupled differential equations in matrix form:

$$\frac{d}{dt}\begin{pmatrix} [\mathrm{A}] \\ [\mathrm{A}^*] \end{pmatrix} = \begin{pmatrix} -k_1[\mathrm{M}] & k_{-1}[\mathrm{M}] \\ k_1[\mathrm{M}] & -(k_{-1}[\mathrm{M}]+k_2) \end{pmatrix}\begin{pmatrix} [\mathrm{A}] \\ [\mathrm{A}^*] \end{pmatrix} \tag{2}$$

The two time constants that govern the kinetics obtained in this solution are the two roots of the characteristic equation:

$$\lambda^2 + (k_1[\mathrm{M}] + k_{-1}[\mathrm{M}] + k_2)\lambda - k_1 k_{-1}[\mathrm{M}]^2 = 0 \tag{3}$$

which are

$$\tfrac{1}{2}(k_2 + (k_1 + k_{-1})[\mathrm{M}]) \pm \tfrac{1}{2}\sqrt{(k_2 + (k_1 + k_{-1})[\mathrm{M}])^2 - 4k_1 k_{-1}[\mathrm{M}]^2} \tag{4}$$

A significant simplification of this solution arises if $k_1 \ll k_{-1}$. This inequality will usually be fulfilled because their ratio is the equilibrium constant for the formation of energized molecules, and for small molecules in thermal equilibrium there will normally be far fewer energized A^* molecules than non-energized A molecules.

Taking a Taylor expansion in powers of k_1 the leading terms for each of the two roots of the quadratic become:

$$\lambda_1 = -\frac{k_2 k_1[\mathrm{M}]}{k_{-1}[\mathrm{M}] + k_2} \tag{5}$$

$$\lambda_2 = -(k_{-1}[\mathrm{M}] + k_2) \tag{6}$$

The first of these time constants is much smaller in magnitude than the second. The predicted kinetics therefore takes place on two timescales: following an initial transient period of fast processes governed by λ_2, there will be a much slower period governed by λ_1, during which the reactant concentration decays away exponentially, and the ratio $[\mathrm{A}^*]/[\mathrm{A}]$ is constant (this ratio is determined by the eigenvector corresponding to λ_1). The faster of these processes is essentially collisional energy transfer leading to the steady-state distribution. The slower process corresponds to the reaction kinetics, which are first-order with respect to A, and so the unimolecular rate coefficient has the value $-\lambda_1$.

This result for the unimolecular rate coefficient may be obtained in a much simpler manner by assuming that, because the energized molecules only ever attain a very low concentration, this concentration obeys the steady-state approximation

$$\frac{d}{dt}[A^*] = k_1[M][A] - (k_{-1}[M] + k_2)[A^*] = 0 \tag{7}$$

implying that

$$[A^*] = \frac{k_1[M][A]}{k_{-1}[M] + k_2} \tag{8}$$

The rate of reaction is

$$k_2[A^*] = \frac{k_2 k_1[M][A]}{(k_{-1}[M] + k_2)} \tag{9}$$

giving the same result as Eq. (6) for the unimolecular rate constant.

This scheme approximates to true first-order kinetics when the rate of collisional deactivation is much greater than the rate of reaction. This follows from Eq. (9) because, if $k_{-1}[M] \gg k_2$, the addition of k_2 in the denominator can be safely neglected, and the rate of reaction becomes equal to

$$\frac{k_2 k_1}{k_{-1}}[A] \tag{10}$$

The physical interpretation of this result is that deactivation is much faster than reaction. Hence, on average, many activations and deactivations will take place before reaction is likely to occur, leading to a pre-equilibrium between A and A^*. The equilibrium constant for this pre-equilibrium is k_1/k_{-1}, and so the rate expression implies in essence that the reaction takes place from an equilibrium concentration of A^* with a rate constant k_2. Because the equilibrium concentration of A^* is proportional to $[A]$ and independent of $[M]$, the kinetics are truly first order and the rate constant is independent of the bath gas concentration $[M]$.

Although the Lindemann theory predicts that the rate constant should be independent of bath gas pressure when $k_{-1}[M] \gg k_2$, this is not the case at lower pressures. The expression for the unimolecular rate coefficient is

$$k_{\text{uni}} = \frac{k_2 k_1}{k_{-1} + k_2/[M]} \tag{11}$$

predicting that the result of Eq. (10) is an upper limit, the 'high pressure limit', and that as the pressure of bath gas is reduced the unimolecular rate coefficient will also decrease. This change of the rate coefficient with pressure of bath gas is known as the 'fall-off' of the rate coefficient. The experimental observation of fall-off was originally used to verify the importance of collisional energization in unimolecular reactions. However, since the general acceptance of the Lindemann mechanism, the fall-off has been regarded as diagnostic of a unimolecular reaction.

If the pressure of bath gas is reduced far enough the inequality $k_{-1}[\mathrm{M}] \ll k_2$ may be attained. In this case the term of $k_{-1}[\mathrm{M}]$ can be ignored in the denominator of Eq. (9) and the rate of reaction becomes equal to $k_1[\mathrm{M}][\mathrm{A}]$.

The physical interpretation of this result is that reaction becomes the dominant process in the competition between reaction and deactivation, because the pressure has become so low that the deactivation process is very slow. Since $k_{-1}[\mathrm{M}] \ll k_2$ and, as discussed above, $k_1 \ll k_{-1}$, it follows that $k_1[\mathrm{M}] \ll k_2$ and that the energized molecules are effectively in a steady state. The activation step becomes rate determining, since almost every energized molecule will react at the first attempt. The rate of reaction is therefore equal to the rate of activation, which is second order overall (first order in reactant and first order in bath gas). This situation is commonly known as the 'low pressure limit'.

The fall-off of the unimolecular rate constant as a function of pressure arises because of the way in which the competition between reaction and deactivation depends on pressure. The rate of reaction is equal to $k_2[\mathrm{A}^*]$, and so its pressure dependence follows the pressure dependence of $[\mathrm{A}^*]$, given by Eq. (8). As the pressure is reduced, the importance of k_2 in the denominator relative to $k_{-1}[\mathrm{M}]$ increases, and so the steady-state concentration $[\mathrm{A}^*]$ is reduced. A useful way of looking at this is that $[\mathrm{A}^*]$ is depleted by reaction, and is lower than its Boltzmann value, $k_1[\mathrm{A}]/k_{-1}$. This reactive depletion is most important at low pressures, where the collisional activation is too slow to replenish it.

The effect of the competition is maximized when the rates of reaction and deactivation are equal, $k_{-1}[\mathrm{M}] = k_2$; at this point the unimolecular rate coefficient is half of k_∞. At the transition pressure

$$k_{\mathrm{uni}} = \tfrac{1}{2}k_\infty = \tfrac{1}{2}k_1[\mathrm{M}]_{\frac{1}{2}} \tag{12}$$

and so $[\mathrm{M}]_{\frac{1}{2}} = k_\infty/k_1$. The 'transition pressure' can be used to estimate k_1. Since $[\mathrm{M}]_{\frac{1}{2}}$ and k_∞ are measurable, k_1 can be calculated from $k_1 = k_\infty/[\mathrm{M}]_{\frac{1}{2}}$.

A more detailed consideration of the fall-off may be obtained by reciprocating Eq. (11), giving

$$\frac{1}{k_{\text{uni}}} = \frac{k_{-1}}{k_2 k_1} + \frac{1}{k_1[\text{M}]} = \frac{1}{k_\infty} + \frac{1}{k_1[\text{M}]} \tag{13}$$

This equation predicts that a graph of $1/k_{\text{uni}}$ against $1/[\text{M}]$ (or $1/p$) will be a straight line whose intercept is $1/k_\infty$ and whose slope is $1/k_1$. This type of plot is called a Lindemann plot, and provides, in principle, an elementary method for determining both the high and low-pressure limiting rate coefficients and for testing the theory. Unfortunately, as discussed below, Lindemann plots are often not linear.

Although the theory does need to be improved in a number of details before it can provide a quantitative description of experiment, the observation of fall-off from first order at high pressures to second order at low pressures is correctly explained by the Lindemann-Christiansen mechanism, and modern theories of unimolecular reactions are based on this mechanism.

2. FAILURES OF THE LINDEMANN THEORY

There are three main respects in which the Lindemann theory needs to be improved.

1. The simple collision theory expression for the activation rate coefficient is in error, often underestimating k_1 by many orders of magnitude.

2. The Lindemann plot described above is typically not a straight line.

3. The idea that a molecule may be activated or deactivated in a single collision is not correct.

These failures will be discussed in more detail in this section. They necessitate more detailed consideration, both of the reaction steps and of the collisional energy transfer steps. The remainder of this work is devoted to these modifications.

2.1. Lindemann-Hinshelwood theory

The first problem discovered with the Lindemann model is the use of the simple collision theory, Eq. (1) for the activation rate constant. The collision frequency Z_{AM} can be calculated simply from collision theory, leaving the activation energy E_0 to be determined. The low pressure limit is often difficult to attain experimentally, because collisions with impurities or collisions between reactant molecules may become important. However, in the expression for the high pressure limit, $k_\infty = k_2 k_1 / k_{-1}$, only k_1 would be expected to have a significant activation energy. k_{-1} should be equal to the collision frequency, which is not activated, and the reaction rate constant k_2 should not have an activation energy, because there is no energy change in the reaction step. It is therefore not unreasonable to estimate the activation energy for k_1 from that of the experimental high pressure limit.

Having done this, it is possible to compare the transition pressure calculated from the theoretical value of k_1 with the experimental value, or equivalently to compare the theoretical value of k_1 with the value inferred from the experimental transition pressure. The problems of this approach are immediately obvious from Fig. 1, which illustrates some data for the decomposition of cyclobutane at 722 K. It is evident that simple collision theory gives a gross underestimate of the low-pressure limiting rate coefficient, and hence the transition pressure estimated from Lindemann theory (1.5×10^8 Torr) is also grossly in error when compared to the experimental value 0.5 Torr. Reference to Fig. 1 shows that it is only possible to put the theoretical and experimental fall-off curves on the same graph using an extended logarithmic scale.

The problem of the assumed form for the activation rate in Lindemann theory is that it is based on the idea that all the energy required for the reaction must be supplied as collisional kinetic energy along the line of centres, thereby neglecting the fact that the molecules have internal structure and already possess both vibrational and rotational energy before the collision. It is less unreasonable than it appears for Lindemann to have assumed this. When Lindemann proposed his theory the new quantum mechanics of Schrödinger and Heisenberg had not been developed and, in consequence, the internal motions of molecules were not well understood. Lindemann believed that the critical form of energy in a unimolecular reaction was rotational energy, and that the molecule dissociated as a result of centrifugal forces, whereas it is now clear that vibrational energy is predominantly important.

The immediate remedy to this problem was to take account of the quantum states of the internal modes of motion (vibration and rotation). Suppose that $\rho(E)\delta E$ is the number of internal quantum states for a molecule in the energy interval from E to $E + \delta E$, and that the probability of activation into this energy range follows the Boltzmann distribution then

$$k_1 = \frac{Z_{\mathrm{AM}}}{Q_{vr}} \int_{E_0}^{\infty} \rho_{vr}(E) \exp(-E/kT) dE \tag{14}$$

where Q_{vr} is the molecular rovibrational partition function. The quantity $\rho(E)$

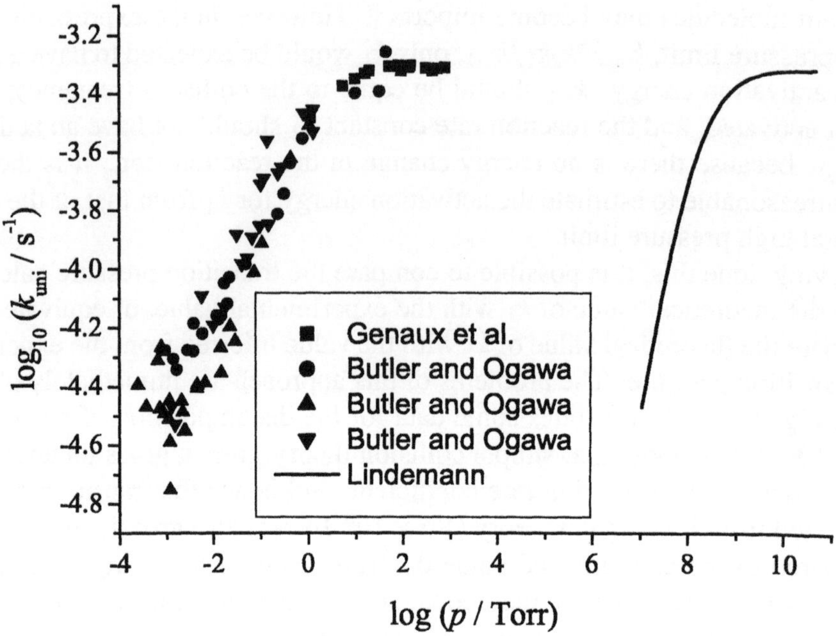

Figure 1. Fall-off data for the dissociation of cyclobutane at 722 K, and comparison with Lindemann theory. Experimental data are from Genaux et al. [6] and Butler and Ogawa [7].

is often referred to as the density of (rovibrational) states. The assumption that the activation rate is proportional to the Boltzmann distribution is equivalent to assuming that the energy of the molecule is completely randomized according to the Boltzmann distribution at every collision, and is known as the *strong collision* approximation.

There is a simple analytical relationship between the density of states and the partition function:

$$Q(T) = \int_0^\infty \rho(E) \exp(-E/kT)\, dE \tag{15}$$

The partition function is the Laplace transform of the density of states, with the Laplace transform variable conjugate to energy being $1/kT$. In principle therefore the density of states may be obtained by finding the inverse Laplace transform of the partition function.

Neglecting rotational motion, and assuming that the vibrational modes are all harmonic, the classical limit of the vibrational partition function of a molecule is

$$Q_v = \prod_{i=1}^{s} \frac{kT}{h\nu_i} \tag{16}$$

where s is the number of vibrational modes, and ν_i are the individual vibrational frequencies. The inverse Laplace transform of this partition function gives the density of states

$$\rho_v(E) = \frac{E^{s-1}}{(s-1)!\prod_{i=1}^{s} h\nu_i} \tag{17}$$

Using this approximation the activation rate coefficient becomes

$$k_1 = Z_{\mathrm{AM}}(1 - P(s, E_0/kT)) \approx \frac{Z_{\mathrm{AM}}}{(s-1)!}\left(\frac{E_0}{kT}\right)^{s-1} \exp(-E_0/kT) \tag{18}$$

where $P(s, E_0/kT)$ is the normalized incomplete gamma function [8]. This expression was first obtained by Hinshelwood [9].

The Hinshelwood formula for the activation rate coefficient is independent of the individual vibrational frequencies, but is sensitive to the number of vibrations. If the full number of vibrations is included in the theory the activation rate coefficient tends to be overestimated by Eq. (18), so that this is an over-correction. It is customary to use s as a variable parameter to fit experimental results. The best fit value for s is typically smaller than (often half) the number of vibrational degrees of freedom.

In the Lindemann-Hinshelwood theory the Lindemann expression for the unimolecular rate constant, Eq. (9), is still assumed to be correct, but an improved activation rate coefficient is obtained from the Hinshelwood formulation. The shape of the fall-off curve should therefore still be the simple form predicted by Lindemann. Reference to Fig. 2 shows that, for the cyclobutane decomposition reaction, the change in the activation rate coefficient brings the theory much closer to the experimental results, particularly at low pressure. However, the shape of the fall-off curve is still not correct: the Lindemann-Hinshelwood model predicts a fall-off region that is too narrow, the true fall-off is broader.

2.2. Curved Lindemann plot

The second problem of the Lindemann-Hinshelwood theory is that the Lindemann plot is often far from a straight line, as can be seen by the example

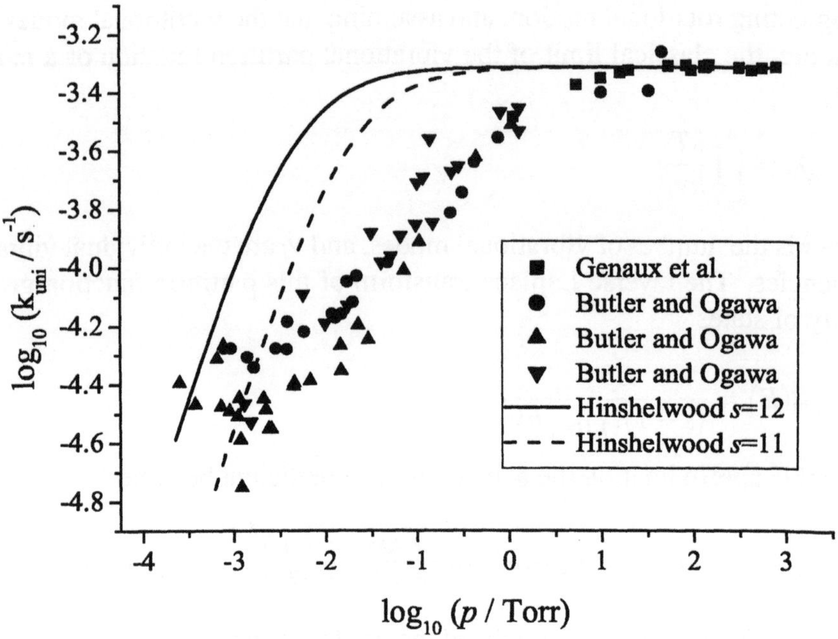

Figure 2. Fall-off data for the dissociation of cyclobutane at 722 K [6,7], and comparison with Lindemann-Hinshelwood theory with 11 or 12 oscillators. The true number of vibrations is 30.

in Fig. 3. As discussed above, the fall-off is a result of the reactive depletion of the population of energized molecules, because of the competition between deactivation and reaction. The lower the pressure, the greater the depletion of the energized molecules, and so the lower the rate coefficient. The main physical reason for the failure of the Lindemann theory to give the correct fall-off is that the rate coefficient for the reaction step k_2 is not the same for all energized molecules.

If the rate constant for the reaction of an energized molecule depends in some way on the quantum state of the reacting molecule, then the degree of reactive depletion of each state will be different for each state, and will depend on the individual rate coefficient for that state.

Although the rate of reaction might be expected to depend on many features of the quantum state of the reacting molecule, it is a reasonable first approxima-

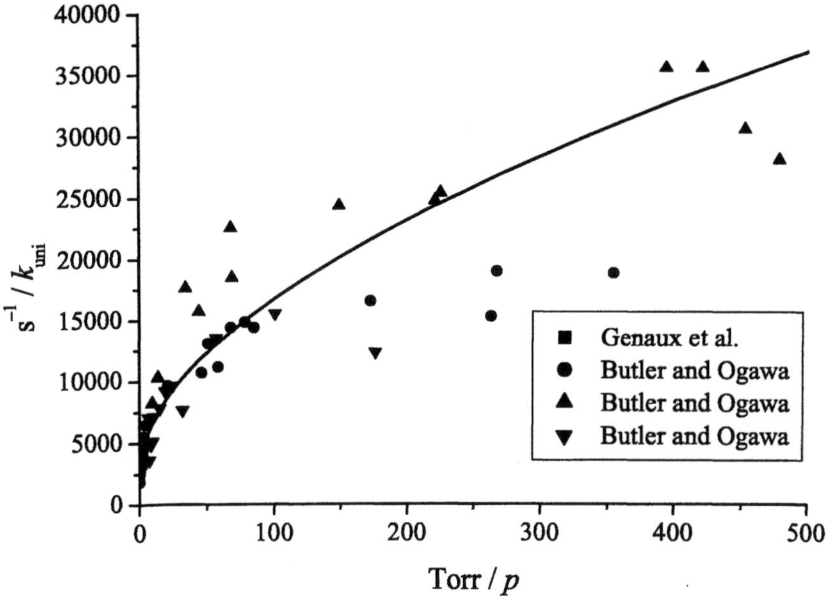

Figure 3. Lindemann plot for the dissociation of cyclobutane at 722 K, illustrating deviation from linearity.

tion to assume that it depends mainly on the energy of the molecule, although this is certainly a simplification: the reaction rate will also depend on angular momentum, and there may be many states with similar energies that react at different rates. However, the minimum acceptable description is that k_2 depends on the energy of the molecule.

If the rate of the reaction step depends on energy it is necessary to consider the model of activation in more detail. The simplest possibility is to follow the strong collision idea of Hinshelwood, that the energy of the molecule is completely randomized according to a Boltzmann distribution at every collision. The rate of activation into the energy interval from E to $E+\delta E$ is therefore

$$\delta k_1(E) = \frac{Z_{AM}}{Q_{vr}}\rho(E)\exp(-E/kT)\delta E \equiv Z_{AM}b(E)\delta E \tag{19}$$

in which $b(E)$ denotes the normalized Boltzmann energy density. Applying this equation, combined with an energy-dependent rate coefficient $k_2(E)$, gives a steady-state concentration of energized molecules in the required energy range

$$\delta\,[\mathrm{A}^*(E)] = \frac{\delta k_1(E)[\mathrm{A}][\mathrm{M}]}{k_{-1}[\mathrm{M}] + k_2(E)} \tag{20}$$

To find the rate of reaction it is necessary firstly to multiply this expression by $k_2(E)$. This procedure gives the contribution to the reaction rate from the energy range δE. Then the expression is integrated over all energies above the reaction threshold, giving

$$k_{\mathrm{uni}} = \int_{E_0}^{\infty} \frac{dk_1(E)[\mathrm{A}][\mathrm{M}]}{k_{-1}[\mathrm{M}] + k_2(E)} = \int_{E_0}^{\infty} \frac{k_2(E) Z_{\mathrm{AM}} b(E)[\mathrm{M}]}{Z_{\mathrm{AM}}[\mathrm{M}] + k_2(E)} dE \tag{21}$$

There is no reason for this expression to be consistent with the Lindemann straight line plot, but it is instructive to examine the physical reasons for the curvature. The low pressure limit is the same as in the Lindemann-Hinshelwood theory because the rate determining step is activation, which is dealt with in the same way in the two theories. This can be seen by taking the low pressure limit of Eq. (21).

$$k_{\mathrm{uni}} = \int_{E_0}^{\infty} Z_{\mathrm{AM}} b(E)[\mathrm{M}] dE \tag{22}$$

This limit is independent of the reaction rate constant $k_2(E)$.

In the high pressure limit there is no reactive depletion, and the population is maintained in a Boltzmann distribution at all energies.

$$k_{\mathrm{uni}} = \int_{E_0}^{\infty} k_2(E) b(E) dE \tag{23}$$

The high pressure limiting rate coefficient is therefore a Boltzmann-weighted average of the energy dependent rate coefficient.

At intermediate pressures the energy dependence of the population distribution is given by

$$f(E) = \frac{Z_{\mathrm{AM}} b(E)[\mathrm{M}]}{Z_{\mathrm{AM}}[\mathrm{M}] + k_2(E)} \tag{24}$$

and the unimolecular rate coefficient is an average over this population distribution:

$$k_{\mathrm{uni}} = \int_{E_0}^{\infty} k_2(E) f(E) dE \tag{25}$$

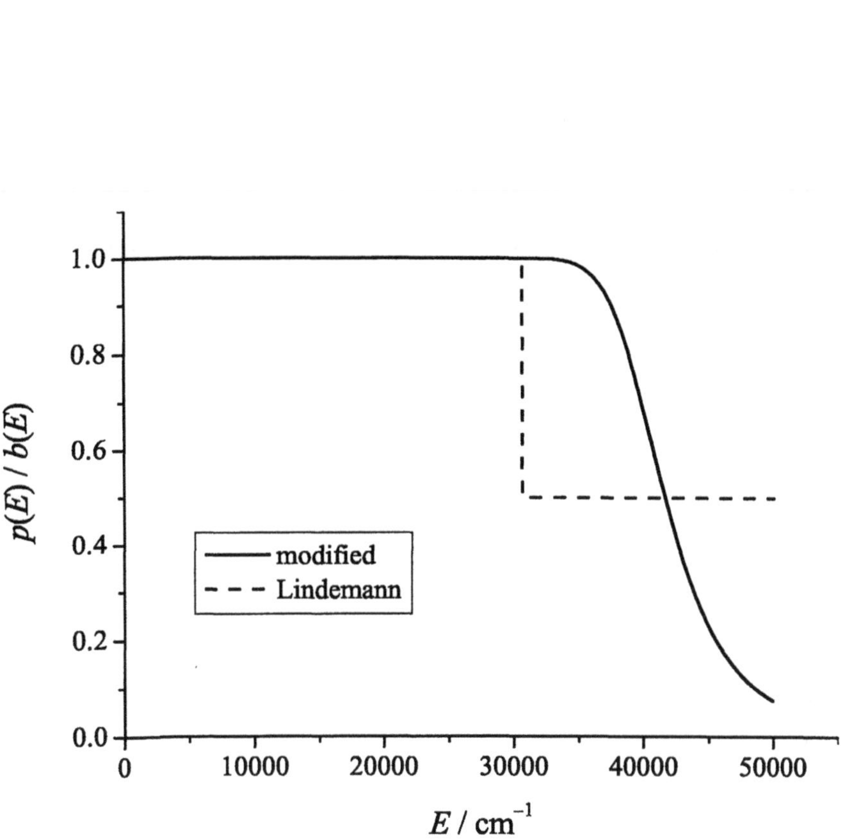

Figure 4. Proportional depletion by reaction with energy-dependent $k_2(E)$ and according to the Lindemann-Hinshelwood model (constant k_2). The curve has been calculated at 999 K for ethane dissociation at the transition pressure.

In the modified model, as illustrated in Fig. 4, the proportional depletion of the Boltzmann distribution depends on the energy. At high energies $k_2(E) \gg k_{-1}[\mathrm{M}]$; these energies are heavily depleted and therefore do not contribute much to the rate. On the other hand, just above the reaction threshold the rate coefficient is small, $k_2(E) \ll k_{-1}[\mathrm{M}]$ and the population is barely depleted at all. These low energy levels contribute almost the full Boltzmann weighting to the rate coefficient, but this is small because $k_2(E)$ is small. Fig. 4 shows the transition between low and high-energy behaviour, which can be thought of as occurring at an energy where $k_2(E) = k_{-1}[\mathrm{M}]$. Clearly the transition energy is an increasing function of the pressure.

In contrast, the Lindemann-Hinshelwood model assumes that all energized molecules react with the same rate constant k_2. This model overestimates the contributions to k_{uni} from high energy states, and underestimates those from low-energy states. The true rate coefficient at some intermediate pressure will fall off faster than predicted by the Lindemann-Hinshelwood theory because of the change in the shape of the depleted population distribution as the pressure is reduced, and this leads to the curvature of the Lindemann plot.

3. RRK THEORY

3.1. Classical RRK theory

It was suggested soon after the development of Lindemann's theory that the critical motions leading to the chemical reaction of a molecule were vibrational rather than rotational. This realization led initially to the simultaneous development of two theories based on very similar assumptions. These theories were proposed by Rice and Ramsperger [10,11] and by Kassel [12]. The theory is based on the idea of Hinshelwood [9] that the molecule is a collection of s classical oscillators, whose density of states given in Eq. (17). One of these oscillators was identified as critical to the reaction, in the sense that the molecule would dissociate immediately if sufficient energy (E_0) were concentrated into this critical oscillator. Such a molecule is said to be activated. The main difference between the Rice-Ramsperger and Kassel theories is that in the former the critical energy is required to be concentrated into one 'square' term, i.e. either the kinetic or the potential energy of an oscillator. This is more restrictive than the Kassel theory, which only requires the energy to be concentrated into an oscillator.

The equilibrium distribution of energy in the molecule is based on the assumption that every state of a given energy is equally probable (the microcanonical distribution). The probability of finding a molecule with energy E in a state where it dissociates is the ratio of the density of states of activated molecules to the total density of states of the molecule. This may be found as follows: suppose that the critical vibration is labelled 1, then the density of states of this critical vibration is given by Eq. (17) with $s = 1$, i.e.

$$\rho_1(E) = \frac{1}{h\nu_1} \tag{26}$$

and the density of states of the remaining $s-1$ vibrations is

$$\rho_{s-1}(E) = \frac{E^{s-2}}{(s-2)!\prod_{i=2}^{s} h\nu_i} \tag{27}$$

For a state to be activated, the energy E must be shared out between the critical coordinate (E_1) and all the others (E_{s-1})in such a way that $E_1 > E_0$. But all values of E_1 greater than E_0 are possible, up to a limit where all the energy of the molecule is in the critical coordinate. Hence the density of activated states is

$$\begin{aligned}\rho_{act}(E) &= \int_{E_0}^{E} \rho_1(E_1)\rho_{s-1}(E-E_1)dE_1 \\ &= \frac{\int_{E_0}^{E}(E-E_1)^{s-2}dE_1}{(s-2)!\prod_{i=1}^{s} h\nu_i} \\ &= \frac{(E-E_0)^{s-1}}{(s-1)!\prod_{i=1}^{s} h\nu_i}\end{aligned} \tag{28}$$

The probability of activation is therefore

$$P_{act} = \left(\frac{E-E_0}{E}\right)^{s-1} \tag{29}$$

The theory then assumes that the rate of reaction is proportional to this probability, and that the microcanonical distribution is not depleted by reaction.

$$k_2(E) = A\left(\frac{E-E_0}{E}\right)^{s-1} \tag{30}$$

It is now straightforward to calculate the high pressure limiting rate coefficient from Eq. (23), giving

$$k_\infty = Ae^{-E_0/kT} \tag{31}$$

The classical RRK theory thus predicted that the high pressure limiting rate coefficient is exactly of the Arrhenius form. In its applications the values of A and E_0 are generally deduced from the experimental temperature dependence of the high pressure limit.

The low pressure limit is identical to that of the Hinshelwood theory because the same expression for the density of states has been used. However, the fall-off is calculated from Eq. (24), which can be written in the form

$$k_{uni} = \frac{Ae^{-E_0/kT}}{(s-1)!}\int_0^\infty \frac{x^{s-1}\exp(-x)}{1+(A/\omega)\left[x/(x+E_0/kT)\right]^{s-1}}dx \tag{32}$$

where ω is the pseudo-first order collision frequency,

$$\omega = Z_{AM}[\mathrm{M}] \tag{33}$$

These Kassel integrals may be evaluated numerically by elementary methods.

3.2. Quantum RRK theory

The classical RRK theory was proposed very soon after the quantum theory of Schrödinger, and so it is scarcely surprising that it uses a classical model of the vibrations. However, very soon afterwards Kassel proposed an alternative version of the RRK theory in which the oscillators were quantum harmonic oscillators [13]. In the simplest version of the theory all the oscillators have the same frequency ν, although Kassel did also present a version in which the oscillators are divided into two classes with different frequencies.

If all the oscillators have the same frequency, the energy of a state simply depends on the number of vibrational quanta $E = nh\nu$, and the density of states in the classical theory is replaced by the degeneracy of this energy level. Since vibrational quanta are indistinguishable (and spinless) the rule for counting their states is the rule that applies to bosons. If there are n quanta in s oscillators, then the degeneracy is equal to

$$g_s(n) = \frac{(n+s-1)!}{n!(s-1)!} \tag{34}$$

This formula may be rationalized by considering the s oscillators as boxes and the n quanta as particles. If the s boxes are arranged along a line, there are $s-1$ partitions between the boxes, each dividing one box from the next. The degeneracy is the number of ways in which $s-1$ partitions can be introduced into a line of n particles. If particles and partitions are considered to be 'things' then there are $n+s-1$ things in total and the degeneracy is the number of ways of choosing n of them to be particles (or equivalently the number of ways of choosing $s-1$ of them to be partitions), leading to the formula of Eq. (34).

The degeneracy can also be obtained by expanding the vibrational partition function

$$Q_\nu = (1 - \exp(-h\nu/kT))^{-s} \tag{35}$$

as a power series in the exponential. The degeneracy $g_n(s)$ is the coefficient of the term $\exp(-nh\nu/kT)$.

The rate coefficient for energization to this energy level is given by the product of the collision frequency and the Boltzmann probability of the level,

$$k_1(n) = \frac{Z_{\mathrm{AM}}}{Q_\nu} g_s(n) \exp(-nh\nu/kT) = \omega g_s(n)(1-\alpha)^s \alpha^n \tag{36}$$

where $\alpha = \exp(-h\nu/kT)$. This is the quantum equivalent of Eq. (19).

In the classical version of the RRK theory a molecule is activated if the energy in a particular oscillator exceeds E_0. In the quantum version this critical energy is identified as an integer number of quanta in the critical oscillator, so that $E_0 = n_0 h\nu$. It is necessary to calculate the degeneracy of the activated states. As before, the total number of vibrational quanta in the molecule (n) must be shared between the critical oscillator and the others. Suppose that there are j quanta in the critical oscillator, according to Eq. (34) the number of ways of distributing the remaining $n-j$ quanta in the other $s-1$ oscillators:

$$g_{s-1}(n-j) = \frac{(n-j+s-2)!}{(n-j)!(s-2)!} \tag{37}$$

The total degeneracy of activated molecules is the sum of this formula over all values of $j \geq n_0$, up to a maximum of n. The summation may be evaluated using a standard recurrence relationship for binomial coefficients [8, Sect. 24.1.1].

$$g_{\text{act}}(n) = \sum_{j=n_0}^{n} \frac{(n-j+s-2)!}{(n-j)!(s-2)!} = \frac{(n-n_0+s-1)!}{(n-n_0)!(s-1)!} \tag{38}$$

The probability of activation is then the ratio of this degeneracy to the total degeneracy,

$$P_{\text{act}} = \frac{(n-n_0+s-1)!n!}{(n-n_0)!(n+s-1)!} \tag{39}$$

As before, the microscopic rate coefficient is assumed to be proportional to this probability, $k_2(n) = AP_{act}$.

The high-pressure limit is

$$k_\infty = Z_{\text{AM}}^{-1} \sum_{n=n_0}^{\infty} k_1(n)k_2(n) = Ae^{-E_0/kT} \tag{40}$$

giving the Arrhenius form, as in the classical RRK theory. This result means that A and E_0 can be obtained from the experimental high pressure limit.

The rate coefficient at lower pressures must be evaluated by the quantum equivalent of the Kassel integral, Eq. (32), which may be written

$$k_{\text{uni}} = Ae^{-E_0/kT}(1-\alpha)^s \sum_{n=0}^{\infty} \frac{g_s(n)\alpha^n}{1+(A/\omega)g_s(n)/g_s(n+n_0)} \tag{41}$$

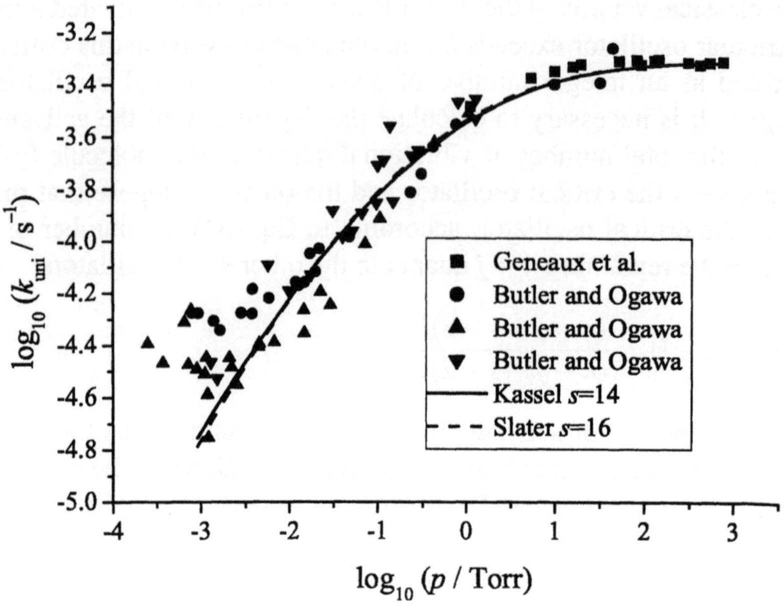

Figure 5. Lindemann plot for the dissociation of cyclobutane at 722 K, compared with the predictions of RRK theory and Slater theory with strong collisions.

3.3. Development of the RRK theories

The RRK theory has the virtue that it is very simple to apply and it does get close to the correct shape of the fall-off curve. As an example, Fig. 5 shows the fall-off curve calculated from classical RRK theory for the dissociation of cyclobutane using 14 oscillators. It can be seen that the theory is a considerable improvement on the Lindemann-Hinshelwood model. There are, however, some remaining problems.

The most obvious difficulty is that the numbers of oscillators in the classical RRK theory required to fit both the low pressure limit and the fall-off are smaller than the actual number of oscillators in the molecule, typically by a factor of about 2, although there can be wide variations [14]. Thus 14 oscillators have been used in the calculation presented in Fig. 5, but cyclobutane actually contains 30 oscillators. The problem is that the use of classical statistical mechanics overestimates the populations of excited states relative to the ground state. For example, in classical theory the mean energy of an oscillator is kT, whereas if $h\nu$ is larger than kT (or comparable with it) most of the population remains in the ground state. This is exactly the problem that leads to the ultraviolet catastrophe in the Rayleigh-Jeans model of the black-body spectrum and motivated the first development of quantum theory.

Several attempts have been made to make s more than just a fitting parameter without losing the simplicity of the model. For example Benson suggested that the number of effective oscillators s could be estimated from the molar vibrational heat capacity, $s = C_{vib}/R$ [15], and Troe has similarly used the energy, $s = U_{vib}/RT$ [16]. However, it seems that no unique value of s is capable of modelling the fall-off curve properly [17], and some attempts have been made to introduce models with two values of s [18].

These problems arise because of the use of the classical density of states rather than the proper vibrational energy levels, and, of course an alternative would be to use the more difficult QRRK theory. If this is done, acceptable fits may be obtained to experimental data using the full number of oscillators. However, QRRK theory is not easily applicable with a realistic spectrum of vibrational frequencies, and it is preferable to use an alternative theory such as the RRKM theory instead.

The other main problem of the RRK theory is that it gives no real insight into the pre-exponential factor A. This parameter has to be obtained from a fit to the high pressure limit. One early attempt to deal with this problem was by Giddings and Eyring [19], who applied ideas of transition state theory to argue that the frequency factor was given by

$$A = \prod_{i=1}^{s} \nu_i \Big/ \prod_{i=1}^{s-1} \nu_i^{+} \tag{42}$$

where ν_i^{+} are the frequencies of the transition state, with the reaction coordinate excluded. This approach is carried through in more detail in the RRKM theory, which will be discussed in Section 5, and more fully in Chapter 2.

4. SLATER THEORY

As early as 1939 Slater proposed an alternative approach that related the rate of a unimolecular reaction to the vibrations of the reacting molecule [55]. This theory was developed over the succeeding years, and is explained in Slater's famous 1959 book [56]. The theory is based on the familiar description of molecular vibrations in terms of normal modes [57]. If the vibrations of a molecule are assumed to be harmonic, they can be reduced to a set of independent harmonic motions, each with its own characteristic frequency and symmetry. The coordinates that oscillate in these motions can be expressed as linear combinations of the atomic displacement coordinates or, more usefully, as linear combinations of a set of internal coordinates, such as stretching of particular bonds or bending of particular bond angles. For example the most useful choice of internal coordinates for the CO_2 molecule comprises the two C – O bond lengths and the OCO bond angle measured in two perpendicular planes whose intersection is the equilibrium configuration of the molecule. Slater's idea was that the reaction is controlled by a particular internal coordinate (such as a bond length) or combination of internal coordinates. This internal motion has components from several of the normal coordinates and so is a superposition of vibrational motions of different frequencies. The reaction takes place when the controlling coordinate reaches some critical value, for example when the bond stretches to a critical length.

The theory of normal coordinates is easier to visualize in terms of the Cartesian displacement coordinates of each atom in the molecule, which span a space of dimension $3N$. Suppose that $\mathbf{r}$ is a column vector of these $3N$ displacements, that $\mathbf{M}$ is a diagonal matrix containing the atomic masses appropriate to each displacement and that $\mathbf{K}$ is a symmetric Hessian matrix containing the force constants, so that the kinetic energy can be expressed $2T = \dot{\mathbf{r}}^T\mathbf{M}\dot{\mathbf{r}}$ (where the dot denotes differentiation with respect to time), and the potential energy is $2V = \mathbf{r}^T\mathbf{K}\mathbf{r}$. The equation of motion for each coordinate is given by Newton's second law with a force equal to minus the gradient of the potential energy:

$$\mathbf{M}\ddot{\mathbf{r}} = -\mathbf{K}\mathbf{r} \tag{43}$$

Although the left hand side is simple because the matrix $\mathbf{M}$ is diagonal, the presence of the matrix $\mathbf{K}$ on the right hand side means that the acceleration of one coordinate must be expressed in terms of a linear combination of the other coordinates. The problem can not be simplified directly by diagonalization of $\mathbf{K}$ because the transformation that diagonalizes $\mathbf{K}$ destroys the diagonal form of $\mathbf{M}$. Geometrically this situation may be visualized as follows: a surface of constant potential energy is an ellipsoid in $3N$ dimensions in the space spanned by $\mathbf{r}$, and a surface of constant kinetic energy is an ellipsoid in the space of the velocities $\dot{\mathbf{r}}$, whose symmetry axes are aligned along the coordinate directions. The transformation of coordinates that diagonalizes $\mathbf{K}$ rotates the coordinates so that the axes are aligned with the symmetry axes of the potential energy ellipsoid, but the kinetic energy ellipsoid is aligned differently, and so the quadratic form expressing the kinetic energy is no longer diagonal.

This situation may be remedied by introducing mass-weighted coordinates,

$$\mathbf{q} = \mathbf{M}^{1/2}\mathbf{r} \tag{44}$$

in which $\mathbf{M}^{1/2}$ is a diagonal matrix whose elements are the square roots of the atomic masses. The kinetic energy is given in terms of these coordinates by $2T = \dot{\mathbf{q}}^T\dot{\mathbf{u}} = \sum_i \dot{\mathbf{q}}_i^2$ and the potential energy is $2V = \mathbf{q}^T\mathbf{f}\mathbf{q}$. (The matrix $\mathbf{f} = \mathbf{M}^{-1/2}\mathbf{K}\mathbf{M}^{-1/2}$ remains symmetric.) Geometrically, the space has been stretched or compressed along each axis direction so that the kinetic energy ellipsoid becomes a $3N$ dimensional sphere. The potential energy surfaces are still ellipsoids in this space. Now the formula for the kinetic energy is invariant to any rotation of the coordinates, and so it is possible to find axes that diagonalize $\mathbf{f}$, while retaining the spherical symmetry of the kinetic energy. If $\mathbf{f}$ is diagonalized by the transformation $\mathbf{Q} = \mathbf{U}^T\mathbf{q}$, where the rows of $\mathbf{U}^T$ are the normalized eigenvectors of $\mathbf{f}$, then the kinetic energy becomes $2T = \sum_i \dot{Q}_i^2$ and the potential energy is $2V = \sum_i \omega_i^2 Q_i^2$, where ω_i^2 is the i'th eigenvalue of $\mathbf{f}$. Each of these *normal coordinates* obeys a separate classical equation of motion,

$$\ddot{Q}_i = -\omega_i^2 Q_i \tag{45}$$

which defines a simple harmonic motion with angular frequency ω_i (frequency $\omega_i/2\pi$), and has the general solution

$$Q_i = A_i \cos(\omega_i t + \phi_i) \tag{46}$$

where A_i is the amplitude of the motion and ϕ_i is the phase. It is important to notice that, because of the mass-weighting, the amplitudes of the normal coordinates are not pure lengths, but have the dimension $M^{1/2}L$. The total energy of normal mode i in this classical model of the molecular vibration is

$$E_i = T_i + V_i = \frac{1}{2}\omega_i^2 A_i^2 \tag{47}$$

and so the amplitude of the normal mode is given by

$$A_i = \frac{\sqrt{2E_i}}{\omega_i} \tag{48}$$

The reason for including this discussion of the normal mode analysis is that Slater uses an unusual convention for the amplitude of the normal modes, in which $A_i = \sqrt{E_i}$, and this convention is also followed in many of the descriptions of his theory in books without explanation. Of course it is perfectly valid to scale the coordinates in any way considered convenient. Slater's scaling (effectively using $Q_i\sqrt{2}/\omega_i$) has the attraction that the surfaces of constant potential energy are spherical, and the surfaces of constant kinetic energy are ellipsoids, rather than the other way round, and that the relationship between the amplitude and the energy is conveniently simple. However, it has been felt desirable to present the formulation in terms of the normal modes more commonly used in vibrational analysis, following the notation of reference [57].

In the quantum mechanical treatment of vibrational normal modes, the vibrational Schrödinger equation is separated into individual harmonic oscillator equations by exactly the same transformation of variables [28].

Six of these normal coordinates (five for a linear molecule) have a frequency eigenvalue identically equal to zero. These motions are translations and rotations of the molecule. Although the approach through Cartesian displacement coordinates is theoretically elegant, it is generally more practical to express the vibrational motions in terms of internal coordinates, such as bond stretches and distortions of bond angles. The method is discussed in detail in Chapter 4 of Wilson, Decius and Cross [57]. Since the distortions of the molecule can be described in terms of $3N-6$ of these internal coordinates there are no redundant dimensions to be removed when the analysis is complete.

The internal coordinates **s** are also expressed as a linear combination of the displacement coordinates. Both the normal coordinates **Q** and the internal coordinates **s** span the same space of $3N-6$ (or $3N-5$) dimensions and so each of the internal coordinates may be written as a linear combination of the normal coordinates:

$$s_i = \sum_j \alpha_{ij} Q_j = \sum_j \alpha_{ij} A_j \cos(\omega_j t + \phi_j) \tag{49}$$

By convention the internal coordinates are expressed as lengths, and so (because of the mass weighting), the coefficients α_{ij} have the dimensions $M^{-1/2}$. The motion of a particular internal coordinate is therefore a superposition of several harmonic motions with different amplitudes and frequencies. This motion is illustrated for the stretching of the N-Cl bond in the molecule NO_2Cl in Fig. 6 using the parameters determined by Slater and tabulated section 3.8 of his book [56].

Since the reaction is deemed to take place when the controlling internal coordinate, s_1, first reaches some critical value, s, from below. This is simplified by assuming that the initial phases are random and finding the average rate of crossing $s_1 = s$ in the upward direction. This problem was solved by Kac [58,59],

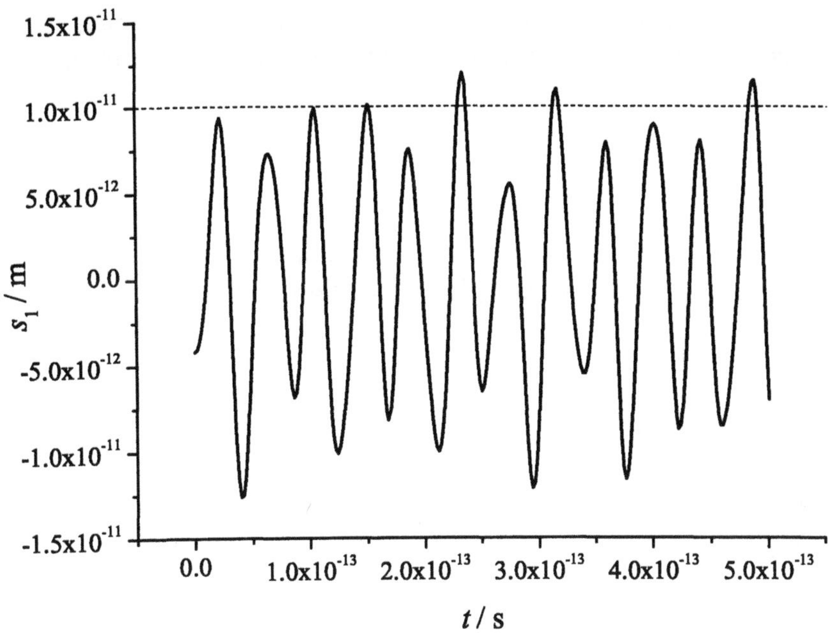

Figure 6. Stretching of the N-Cl bond from its equilibrium value in NO_2Cl. Energy equivalent to 1 quantum has been placed in each of the three contributing normal modes.

and the solution is given by the equation

$$L = \pi^{-2} \int_0^{\infty} \int_0^{\infty} \frac{\cos(sx)}{y^2} \times \left[\prod_j J_0(\alpha_{1j} A_j x) - \prod_j J_0\left(\alpha_{1j} A_j \sqrt{x^2 + \omega_j^2 y^2}\right) \right] dx\, dy \quad (50)$$

where J_0 is a Bessel function.

This equation is complicated to deal with; in particular, because of the relationship between the amplitude of each normal mode and the energy in that mode, it gives the rate in terms of the amount of energy in each of the contributing normal modes. It is useful to consider some simple assumptions about this energy distribution.

Firstly it is necessary to find the minimum energy for which reaction can take place. Molecules can only react if the maximum value of the controlling internal coordinate is greater than the required critical value, i.e.

$$s_{\max} = \sum_j |\alpha_{1j}| A_j \geq s \tag{51}$$

The lowest total energy in the important vibrations for which this inequality is obeyed is easily found (for example by the method of Lagrange multipliers) to be

$$E_0 = \frac{s^2}{\sum_j 2\alpha_{1j}^2/\omega_j^2} \tag{52}$$

Defining the dimensionless parameter μ_j by

$$\mu_j = \frac{|\alpha_{1j}|/\omega_j}{\left(\sum_j \alpha_{1j}^2/\omega_j^2\right)^{1/2}} \tag{53}$$

then the critical minimum energy distribution in the contributing normal modes is given by

$$E_{j0} = \sum_j \mu_j^2 E_0 \tag{54}$$

In the high-pressure limit the energy in each mode is Boltzmann distributed. The high-pressure canonical rate constant can then be found by averaging the rate of Eq. (50) over the Boltzmann distribution. The integrations are not elementary, but are detailed by Slater [56, Chapter 5] and yield the result

$$k_\infty = \nu e^{-E_0/kT} \tag{55}$$

where $2\pi\nu = \left(\sum_j \alpha_{1j}^2 / \sum_j 2\alpha_{1j}^2/\omega_j^2\right)^{1/2}$, which is a kind of average of the contributing normal mode frequencies.

In a similar way an average rate coefficient for molecules with total energy E may be obtained and is found to be

$$k_2(E) = \nu \left(\frac{E - E_0}{E}\right)^{n-1} \tag{56}$$

for energies greater than E_0, where n is the number of contributing vibrational modes. However, it is important to realise that this rate constant is an approximation of Eq. (50) because it assumes that all states with the same energy are equally probable (the microcanonical distribution). Because the reaction rate depends on the detailed energy distribution, different states of the same energy will be depleted differently by reaction and therefore have different steady-state populations.

Although it would be possible to use the approximate microcanonical rate coefficient in the general framework for pressure-dependent rate coefficients, Eq. (24), it is an improvement to use the more detailed Slater model. Unfortunately the integrals required using Eq. (50) are too complicated, and so Slater developed a more tractable approximate form, which is valid when the critical value of the controlling coordinate is close to the maximum possible value for the given energy.

$$L \approx \frac{1}{\Gamma((n+1)/2)} \left[\frac{\sum_j |\alpha_{1j}|A_j v_j^2}{\prod_j |\alpha_{1j} A_j} \right]^{1/2} \left(\frac{s_{\max} - s}{2\pi} \right)^{(n-1)/2} \tag{57}$$

The depleted population of a particular state is given by Eq. (24) with $k_2(E)$ replaced by the more detailed L. The approximate value of the unimolecular rate coefficient is then given by

$$k_{\text{uni}} = \frac{k_\infty}{m!} \int_0^\infty \frac{x^m e^{-x}}{1 + x^m/\theta} dx \tag{58}$$

where $m = (n-1)/2$, $\theta = (\omega/v) f_m b^m$, $b = E_0/kT$ and $f_m = m!(4\pi)^m \mu_1 \ldots \mu_n$. (The quantities μ_j are defined in Eq. (53)).

This concludes the discussion of early theories of the reaction step. These have largely been superseded by RRKM theory and the Statistical Adiabatic Channel Model, which are discussed in the next two sections.

5. RRKM THEORY

RRKM theory, developed from RRK theory by Marcus and others [20–23], is the most commonly applied theory for microcanonical rate coefficients, and is essentially the formulation of transition state theory for isolated molecules. An isolated molecule has two important conserved quantities, 'constants of the motion', namely its energy and its angular momentum. The RRKM rate coefficient for a unimolecular reaction may depend on both of these. For the sake of simplicity this treatment will initially only consider the energy: angular momentum effects will be considered later. Chapter 2 is devoted to modern aspects of the application of RRKM theory to real reactions.

As in the development of RRK theory, the analysis starts with the assumption that the dynamics of the molecule are classical. The essential approximations of the classical RRKM treatment are that

1. The reaction takes place on a single potential energy surface (i.e. it is adiabatic).

2. The motion on that surface is classical.

3. There is a unique reaction coordinate that links reactants to products along the surface, and is separable from the other molecular motions.

4. There is a unique dividing surface, the transition state, dividing reactants from products, which is a point of no return, in the sense that any trajectory passing forwards through it must reach products without turning back.

5. The distribution of reactants is defined by a microcanonical ensemble.

This set of approximations is essentially similar to that for the more familiar canonical transition state theory, apart from the final assumption, that the state is described by a microcanonical distribution (i.e. fixed energy) rather than the Boltzmann distribution (thermal equilibrium) of canonical transition state theory.

Of course, all of these assumptions are open to question:

- Many reactions are non-adiabatic and involve transitions between surfaces.

- Molecular dynamics are properly quantum-mechanical, so that tunnelling may play a part, for example. The quantum view of reactions is considered in Chapter 3.

- Quantum properties, such as tunnelling, do not necessarily follow the reaction coordinate.

- Some trajectories pass through the transition state and return to the reactants without reaching products. Others may recross the transition state several times before attaining products.

- The microcanonical ensemble may be depleted in the vicinity of the transition state by the absence of trajectories in the reverse direction. This assumption is often referred to as the *ergodic* approximation, that the microcanonical ensemble is rapidly randomized behind the reaction bottleneck faster that reactive loss can perturb the distribution.

Many of these shortcomings can be corrected, as discussed in Chapter 2, ultimately giving a powerful and successful theory. However, for the present this section concentrates on the basic principles.

5.1. Classical RRKM theory

The potential energy surface for a reaction is a function of the $3N-6$ coordinates that determine the relative positions of the atoms of the system ($3N-5$ for a linear system). A particular configuration of the atoms is represented by a point on the potential energy surface, and as the atoms move during the reaction, so the point follows a trajectory over the surface, obeying the laws of classical (Newtonian) mechanics. A simple view of the classical rate of reaction is the net rate at which these trajectories pass from reactants to products. This idea is made more rigorous in the approach by Nordholm [24], discussed in the book by Gilbert and Smith [25].

The key idea that supplements RRK theory is the transition state assumption. The transition state is assumed to be a point of no return. In other words, any trajectory that passes through the transition state in the forward direction will proceed to products without recrossing in the reverse direction. This assumption permits the identification of the reaction rate with the rate at which classical trajectories pass through the transition state. In combination with the ergodic approximation this means that the reaction rate coefficient can be calculated from the rate at which trajectories, sampled from a microcanonical ensemble in the reactants, cross the barrier, divided by the total number of states in the ensemble at the required energy. This quantity is conveniently formulated using the idea of *phase space*.

5.1.1. Phase space

In classical mechanics the state of a system is completely defined by specifying the positions and momenta of all the particles simultaneously. This means that 6 coordinates (3 position coordinates and 3 momentum coordinates) must be specified for each particle. The $6N$ dimensional space spanned by the coordinates required for a system of N particles is called *phase space*.

The instantaneous state of the system at a given moment is defined by the values of the $6N$ coordinates, i.e. it constitutes a point in phase space. The dynamical evolution of the system is described by the trajectory of that phase point.

Classical statistical mechanics is concerned with the probability distribution of phase points. In a classical microcanonical ensemble the phase space density is constant. Loosely speaking, all phase points with the same energy are equally likely. In consequence the number of states of the classical system in a given energy range E to $E+dE$ is proportional to the volume of the phase space shell defined by this energy range.

In order to make the theory useful it is necessary to know the constant of proportionality, which is calculated in such a way as to give the classical limit of the number of quantum states. This matter is dealt with in standard books on statistical mechanics [26]. The result is that for a system with n degrees of freedom, i.e. n position coordinates $\mathbf{q}$ and n momentum coordinates $\mathbf{p}$, the number of states in the infinitesimal volume element $d\mathbf{q}\,d\mathbf{p}$ is equal to $d\mathbf{q}\,d\mathbf{p}/h^n$, where n is Planck's constant. The association of a phase space volume h^n with each quantum state can be thought of as a consequence of the uncertainty principle, which limits the precision with which a phase point can be specified in a quantum mechanical system.

Throughout the evolution of the trajectory three quantities remain constant: the energy, denoted $H(\mathbf{q},\mathbf{p})$, the overall angular momentum vector $\mathbf{L}(\mathbf{q},\mathbf{p})$ and the linear momentum of the whole system. The last of these is of no consequence for chemical reactions, because the inertial motion of the centre of mass can be rigorously separated from the internal motions of the system, but the other two quantities need more careful consideration.

5.1.2. The density of states

The density of states is a central concept in the development and application of RRKM theory. The density of states of a classical system is the number of states of that system per unit energy, expressed as a function of energy. This quantity may be formulated as a phase space integral in several ways.

The most natural approach to the density of states is to find the volume of phase space corresponding to the shell defined by the energy range $E \leq H(\mathbf{q},\mathbf{p}) < E+\varepsilon$, and divide this by the span of the energy range ε before taking the limit $\varepsilon \to 0$.

$$\rho(E) = h^{-n} \lim_{\varepsilon\to 0} \varepsilon^{-1} \int \mathbf{1}_{(E,E+\varepsilon)}(H(\mathbf{q},\mathbf{p}))\, d\mathbf{q}\, d\mathbf{p} \tag{59}$$

in which $\mathbf{1}$ is an indicator function, which is unity when its argument lies within the specified interval and zero elsewhere.

It is easy to see that the definition above is essentially a differential. It is therefore entirely equivalent to find the total number of states up to and including E by a suitable phase space integral and then differentiate with respect to E.

$$\rho(E) = h^{-n}\frac{d}{dE}\int \mathbf{1}_{(0,E)}(H(\mathbf{q},\mathbf{p}))\, d\mathbf{q}\, d\mathbf{p} \tag{60}$$

It is sometimes convenient to do the same thing by using a Dirac delta function to project out the required value of the energy, for example

$$\rho(E) = h^{-n}\int \delta(H(\mathbf{q},\mathbf{p}) - E)\, d\mathbf{q}\, d\mathbf{p} \tag{61}$$

From this last definition it is easy to see that the density of states measures the number of states of the system per unit energy at the given energy. In the development of RRKM theory the density of states of the reactant molecule is required, and so this integral must be evaluated over that part of phase space that lies on the reactant side of the dividing surface.

5.1.3. Rate of flow through the transition state

It is now necessary to separate the reaction coordinate, denoted q_1, from the other motions of the system, which will be denoted $\mathbf{q}^+$. At given values of the reaction coordinate q_1 and its conjugate momentum p_1 the kinetic energy due to motion along the reaction coordinate is $p_1^2/2\mu_1$ and the contribution of the reaction coordinate to the potential energy can be written $V(q_1)$. The quantity μ_1 is the reduced mass for motion along the reaction coordinate. Hence the energy can be divided

$$E = E^+ + V(q_1) + \frac{p_1^2}{2\mu_1} \tag{62}$$

where E^+ is the energy of the motions normal to the reaction coordinate.

Now this separation can be made at the transition state,

$$E = E^+ + E_0 + \frac{p_1^2}{2\mu_1} = E^+ + E_0 + E_1 \tag{63}$$

where the potential energy in the reaction coordinate is denoted E_0, and E_1 is the excess energy (above E_0) in the reaction coordinate, i.e. the kinetic energy in the reaction coordinate at the transition state.

To work out the rate at which states of a given energy pass through the transition state, $R^\dagger$ (per unit energy) one can assume that the transition state is defined by an infinitesimal interval $q^\dagger \leq q_1 < q^\dagger + \varepsilon$, which can be indicated by the indicator function $\mathbf{1}_{(q^\dagger, q^\dagger+\varepsilon)}(q_1)$. If the reaction coordinate is travelling with a speed v_1, it takes time ε/v_1 to pass through the transition state, so that the rate of passing through the transition state is $v_1/\varepsilon = p_1/\mu_1\varepsilon$. Hence

$$R^\dagger = h^{-n} \lim_{\varepsilon \to 0} \int \delta(H(\mathbf{q},\mathbf{p}) - E)\mathbf{1}_{(q^\dagger, q^\dagger+\varepsilon)}(q_1)\frac{p_1}{\mu_1 \varepsilon} d\mathbf{q}\, d\mathbf{p} \tag{64}$$

Alternatively, and more conveniently, the factor ε^{-1} can be subsumed into the indicator function and the order of the integration and the limiting process interchanged, introducing another delta function:

$$R^\dagger = h^{-n} \int \delta(H(\mathbf{q},\mathbf{p}) - E)\delta(q_1 - q^\dagger)\frac{p_1}{\mu_1} d\mathbf{q}\, d\mathbf{p} \tag{65}$$

To evaluate this rate the reaction coordinate variables are separated out,

$$R^\dagger = h^{-n} \int \delta(H(\mathbf{q},\mathbf{p}) - E) d\mathbf{q}^+\, d\mathbf{p}^+ \delta(q_1 - q^\dagger)\frac{p_1}{\mu_1} dq_1 dp_1 \tag{66}$$

The inner integrals must be performed as functions of q_1 and p_1, giving

$$R^\dagger = h^{-n} \int \delta(H(\mathbf{q}^+, \mathbf{p}^+) + V(q_1) + \frac{p_1^2}{2\mu_1} - E)\, d\mathbf{q}^+\, d\mathbf{p}^+ \times \delta(q_1 - q^\dagger)\frac{p_1}{\mu_1} dq_1 dp_1 \tag{67}$$

in which the inner integrals are seen by reference to Eq. (61) to represent the density of states for the motions excluding the reaction coordinate, denoted ρ^+.

$$R^\dagger = h^{-1} \int \rho^+(E - V(q_1) - \frac{p_1^2}{2\mu_1})\delta(q_1 - q^\dagger)\frac{p_1}{\mu_1}\, dq_1 dp_1 \tag{68}$$

It is straightforward to integrate out the reaction coordinate, setting $V(q^\dagger) = E_0$ and $p_1^2/2\mu_1 = E_1$,

$$R^\dagger = h^{-1} \int \rho^+(E - E_0 - E_1)\frac{p_1}{\mu_1} dp_1 \tag{69}$$

Finally the change of variable $E_1 = p_1^2/2\mu_1$ gives

$$R^\dagger = h^{-1}\int_0^{E-E_0} \rho^+(E-E_0-E_1)dE_1 \tag{70}$$

The limits on the integral indicate that there must be some kinetic energy in the reaction coordinate, but there may not be more than $E-E_0$ because E_0 must be taken up by the potential energy of the reaction coordinate at the transition state.

A further change of variable $E^+ = E - E_0 - E_1$ gives the simple form

$$R^\dagger = h^{-1}\int_0^{E-E_0} \rho^+(E^+)dE^+ \tag{71}$$

5.1.4. RRKM rate coefficient

The RRKM microcanonical rate coefficient is the rate at which states pass through the transition state (per unit energy) divided by the total density of states of the reactants.

$$k(E) = \frac{\int_0^{E-E_0} \rho^+(E^+)dE^+}{h\rho(E)} \tag{72}$$

The numerator of this expression is the number of states at the dividing surface that have energy between 0 and $E-E_0$ in the motions orthogonal to the reaction coordinate. This quantity is often called the sum of states, and given the symbol W^+.

$$k(E) = \frac{W^+(E-E_0)}{h\rho(E)} \tag{73}$$

5.2. Improvements to RRKM theory

This analysis has been very simple, and several improvements need to be made before the theory is usable.

5.2.1. Fixed and non-fixed energy

In the development above it has been implicitly understood that all motions of the molecule orthogonal to the reaction coordinate apart from the centre of mass motion are included in ρ^+, and that all of the energy of the molecule above the barrier height is available to be transferred to the reaction coordinate at the transition state, and therefore contributes to the rate coefficient.

These ideas are both oversimplifications. The molecule clearly has energy that does not contribute to the reaction rate. The centre of mass motion has already been separated out in this respect, but some other types of energy can not contribute to the reaction rate either. For example some of the vibrations remain more-or-less unchanged from reactants to products (conserved modes), the zero point energy of these molecular vibrations is always present and can not be redistributed into the kinetic energy of the reaction coordinate. This quantity must therefore be excluded from both E and E_0, which should both be considered to be classical energy minus zero point energy. Often the zero-point energy of the transition state is similar to that for the reactant, so that this has little effect on $E - E_0$, but this is not necessarily the case. If the zero-point energy changes through the course of the reaction, the rate will be affected.

Rotational energy is a more difficult problem. Rotational and vibrational motions can couple. An obvious example is the possibility that centrifugal forces in a rotating molecule may contribute to bond breaking. In addition there is Coriolis coupling that permits the flow of energy between rotation and vibration, although this is a weaker effect. All of this is constrained by the conservation of the total angular momentum. If a rotational motion stays in the same quantum state throughout the reaction then it is essentially a spectator and its energy does not contribute to the reaction rate. Such a degree of freedom is said to be *adiabatic*, and can be excluded from the calculation. However, it should be remembered that the classical rotational energy is equal to

$$E_r = \frac{L_x^2}{2I_{xx}} + \frac{L_y^2}{2I_{yy}} + \frac{L_z^2}{2I_{zz}} \tag{74}$$

where the angular momentum vector is $\mathbf{L}$ and the three principal moments of inertia are I_{xx}, I_{yy} and I_{zz}. The angular momentum vector is conserved, but if the moments of inertia change during the course of the reaction then the rotational energy will change and some of this can contribute to the reaction rate, as for the zero-point energy above. A fairly normal practice is to assume that one of the rotational modes is active, i.e. its energy is non-fixed, whereas the other two are approximately adiabatic. This idea is discussed in detail in the book by Forst [27].

In most molecules that have sufficient energy to react the anharmonicities of the vibrations cause such rapid mixing that the vibrational energy of the molecule can be assumed to randomize rapidly on the timescale of the reaction. However, particularly for small molecules, there may be spectator adiabatic vibrational modes, whose coupling with the reaction coordinates is very weak. An example of an adiabatic vibration is discussed in depth in Chapter 3.

5.2.2. Angular momentum conservation

The main effect of rotation on unimolecular reaction rates is the centrifugal effect mentioned above. The origin of this effect can be demonstrated clearly for a rotating diatomic molecule whose potential energy curve is $V(r)$. When the centre of mass coordinates have been separated out there are three remaining degrees of freedom, the bond length r and two angles that determine the orientation of the molecule in space, θ and ϕ.

The classical energy of the system can be expressed in Cartesian coordinates and is simply transformed into spherical polar coordinates,

$$E = \frac{\mu}{2}(\dot{x}^2+\dot{y}^2+\dot{z}^2)+V(r) \tag{75}$$

$$= \frac{\mu}{2}(\dot{r}^2+r^2\dot{\theta}^2+r^2\sin^2\theta\dot{\phi}^2) \tag{76}$$

where μ is the reduced mass and the dots denote differentiation with respect to time. The angular momentum vector is also a constant of the motion, and its square can also be expressed in spherical polar coordinates as

$$L^2 = \mu^2 r^4(\dot{\theta}^2+\sin^2\theta\dot{\phi}^2) \tag{77}$$

Comparing these two equations, and defining a vibrational momentum $p_r = \mu\dot{r}$, the energy takes the suggestive form

$$E = \frac{p_r^2}{2\mu}+V(r)+\frac{L^2}{2\mu r^2} \tag{78}$$

The first term can be interpreted as the vibrational kinetic energy, and the second and third terms are the effective potential energy for the vibrational motion. The final term, which is the rotational kinetic energy, becomes an addition to the potential energy in the non-inertial frame of reference of the rotating molecule.

The same conclusion can be found for a quantum mechanical model of a rotating diatomic molecule by separation of angular and radial variables in the corresponding Schrödinger equation [28]. The centrifugal energy is a decreasing function of the bond length. Its effect is to reduce the effective dissociation energy, although at long distances it becomes the dominant effect and a centrifugal barrier is induced. The increase of the centrifugal barrier with angular momentum is illustrated in Fig. 7.

Although the centrifugal barrier is important, the main effect is the release of rotational energy into the changing bond length (the centrifugal force). The basic RRKM result can be generalized for the constraint of angular momentum

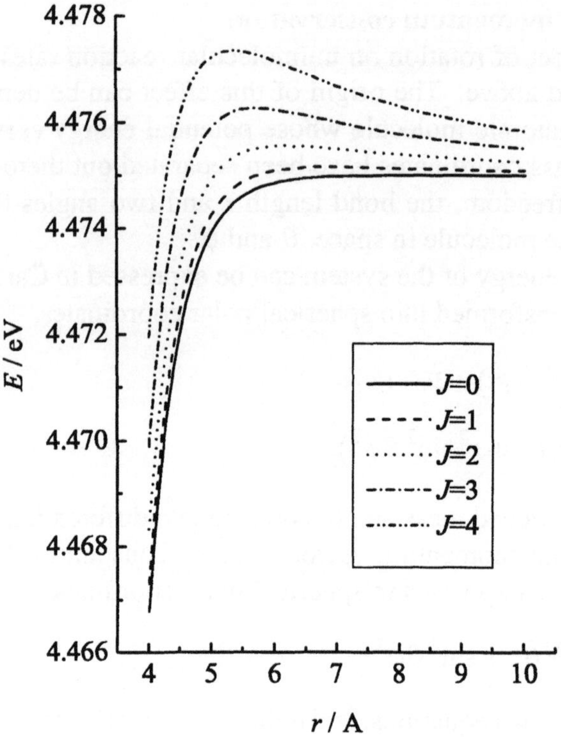

Figure 7. Development of a centrifugal barrier at large separations for the H_2 molecule.

conservation as follows. Suppose that the rotational energy of the reactants is E_{rot} and that of the transition state, given the constraint of angular momentum conservation, is $E^{\dagger}_{rot}$. Then the extra energy available at the transition state is $\Delta E_{rot} = E^{\dagger}_{rot} - E_{rot}$, which is effectively a correction to the barrier height:

$$k(E,J) = \frac{\int_0^{E+E_{rot}-(E_0+E_{rot})^{\dagger}} \rho^{+}(E^{+},J)dE^{+}}{h\rho(E,J)} \tag{79}$$

The derivation of this result is discussed in more detail in the book by Gilbert and Smith [25].

5.2.3. Statistical factor

The derivation of RRKM theory given here is also deficient because it has assumed that there is a single unique reaction path between the reactants and the products. In fact there may be several such paths related by symmetry, each with its own transition state. For example in the dissociation of CH_4 to $CH_3 + H$ there are obviously four equivalent paths. The correct way to allow for this reaction path degeneracy has been the subject of much controversy [25,27,29–34]. However this was finally resolved by Pechukas [35], and the resolution of the problem is relatively simple.

The main effect is already taken into account if symmetry numbers are included in the densities of states. The symmetry number is a correction to the density of states that allows for the fact that indistinguishable atoms occupy symmetry-related positions and these atoms have to obey the constraints of the Pauli principle (i.e. the wave function must have a definite symmetry with respect to any permutation), whereas the classical density of states contains no such constraint. The density of states is reduced by a factor that is equal to the dimension of the rotational subgroup of the molecule. When a molecule is distorted, its symmetry is reduced, and so its symmetry number changes by a proportion that is equivalent to the number of indistinguishable ways in which the distortion may be produced. For example, the rotational subgroup of the methane molecule is T, whose dimension is 12, whereas the rotational subgroup of a distorted molecule in which one bond is stretched is C_3, whose dimension is 3. The ratio of these symmetry numbers, 4, is the number of ways in which the distortion can occur, i.e. the reaction path degeneracy.

The symmetry numbers also allow for the degeneracy of possible return paths from the transition state, which has caused problems in the past [34,32]. The only corrections necessary to the use of symmetry numbers are the additional possibility that the reactant or the transition state may have a number of optical isomers. In this case Gold has shown that the rate coefficient needs to be corrected by a factor $m^\dagger/m$ equal to the ratio of the number of optical isomers in the transition state to that for the reactant [31].

5.2.4. Variational methods

One of the explicit assumptions of RRKM theory, and of transition state theory, is that of a unique dividing surface on the reaction path, which acts as a point of no return. This assumption leads to the conclusion that the rate of reaction can be equated to the rate at which trajectories cross this dividing surface in the direction of products. This assumption is not correct – it has been known for some time that some trajectories return after passing through the transition state and may recross later. Such trajectories are counted more than once and lead to an overestimate of the rate coefficient.

The fact that the RRKM rate coefficient is an upper bound on the exact classical solution for the rate coefficient is actually very useful, as it leads to a method for optimizing the position of the transition state. It is not always obvious where on the reaction coordinate the transition state should be located. If the reaction has a well defined potential barrier the natural place to locate the transition state is at the maximum of the potential barrier (the transition structure). However, if there is no barrier, there is no obvious location for the transition state. Since it is guaranteed that the theory will overestimate the classical rate coefficient, the position of the transition state along the reaction coordinate can be varied until the calculated rate coefficient passes through a minimum. This minimum value will be the optimal estimate of the rate coefficient and the corresponding location of the transition state will also be optimal.

Variational transition state theory was suggested by Keck [36] and developed by Truhlar and others [37,38]. Although this method was originally applied to canonical transition state theory, for which there is a unique optimal transition state, it can be applied in a much more detailed way to RRKM theory, in which the transition state can be separately optimized for each energy and angular momentum [37,39,40]. This form of variational microcanonical transition state theory is discussed at length in Chapter 2, where there is also a discussion of the variational optimization of the reaction coordinate.

5.2.5. Density of states

Calculation of the RRKM rate coefficient requires knowledge of the density of states of the reactants and the transition state, and a great deal of effort has been expended in working out methods for calculating, simulating or estimating these quantities.

Because of its classical nature, RRKM theory has the some of the same defects as the classical RRK theory, discussed in section 3 above. Thus, when the density of states and the sum of states are calculated it is necessary to take account of the fact that these are actually quantum states and not a continuum.

Methods for calculating these quantities are discussed in detail in the book by Holbrook, Pilling and Robertson [29]. These methods fall into several basic categories – classical approximations, inversion of the partition function, direct count methods and Monte Carlo methods, each of which is introduced briefly.

Firstly, it is convenient to assume that rotational motion and vibrational motion are approximately separable from one another, notwithstanding the discussion of angular momentum conservation earlier in this section. The energy of the molecule E is therefore approximately the sum of the vibrational energy and the rotational energy. If the rotational energy is E_r and the density of rotational states at this energy is $\rho_r(E_r)$, then the vibrational energy is $E - E_r$ and the density of vibrational states at this energy is $\rho_v(E - E_r)$. Each of the possible rotational states of energy E_r may be combined with any of the possible vibrational states of energy $E - E_r$, and so the total contribution to the density of states from molecules whose rotational energies are in the range $(E_r, E_r + dE_r)$ is therefore $\rho_r(E_r)\rho_v(E - E_r)\, dE_r$. Integrating over all possible rotational energies

$$\rho(E) = \int_0^E \rho_r(E_r)\rho_v(E - E_r)\, dE_r \tag{80}$$

This integral is a convolution, which is equivalent to the separation of the rovibrational partition function into the product of separate rotational and vibrational partition functions, by the convolution theorem of Laplace transforms.

The classical approximation for the rotational density of states of a molecule is familiar from elementary statistical mechanics, where it is common to assume that the rotational states form a continuum in calculating the rotational partition function. For the external rotations of most molecules this approximation is very good. For example the classical approximation for the rotational partition function of an asymmetric top is

$$Q_r = \frac{1}{\sigma}\left(\frac{2\pi kT}{\hbar^2}\right)^{3/2} \sqrt{I_{xx}I_{yy}I_{zz}} \tag{81}$$

where σ is the symmetry number. This partition function is related to the rotational density of states by a Laplace transform

$$Q_r(\beta) = \int_0^\infty \rho_r(E_r)\exp(-\beta E_r)\, dE_r \tag{82}$$

and so the rotational density of states may be obtained by inverting this Laplace transform, in which $\beta = 1/kT$ is the Laplace transform variable conjugate to E_r. The inversion gives

$$\rho_r(E_r) = \frac{4\pi\sqrt{I_{xx}I_{yy}I_{zz}}}{\hbar^3}\sqrt{2E_r} \tag{83}$$

Similar methods may be used for individual rotors, for symmetric tops and for systems containing internal rotors, as long as the exact partition function is known. There are, however, problems associated with internal rotation, because the rotations may be hindered, and the internal rotors are not independent because of the conservation of angular momentum. The resolution of these problems has been discussed by Robertson and Wardlaw [41], and is dealt with in Chapter 2.

Although the classical approximation is generally acceptable for rotations the same is not true for vibrations. The classical density of states for the vibrations of a molecule consisting of s oscillators, all assumed harmonic was given in Eq. (17).

$$\rho_v(E_v) = \frac{E_v^{s-1}}{(s-1)!\prod_{i=1}^{s} h\nu_i} \tag{84}$$

Comparison with direct count methods for systems of harmonic oscillators indicates that the classical density of states is generally an underestimate. However, it is so convenient to have an explicit formula for the density of states that more accurate semiclassical approximations have been developed.

Marcus and Rice [20] suggested that the reason for the underestimate was that the vibrational energy of a molecule, E_v, was measured relative to the zero point level and not the equilibrium geometry (the minimum of the potential well), and so the density of states could be corrected by adding the vibrational zero-point energy to E_v,

$$\rho_v(E_v) = \frac{(E+E_z)^{s-1}}{(s-1)!\prod_{i=1}^{s} h\nu_i} \tag{85}$$

This corrected approximation is a great improvement on the uncorrected classical method, but is an overcorrection, leading to a systematic overestimate of the density of states. Whitten and Rabinovitch [42] corrected this overestimate by introducing a parameter a to weight the zero-point energy in the correction formula.

$$\rho_v(E_v) = \frac{(E+aE_z)^{s-1}}{(s-1)!\prod_{i=1}^{s} h\nu_i} \tag{86}$$

The value of the parameter a, which lies between 0 and 1, must be determined for each system by comparison with more accurate methods, but once this has been done the Whitten-Rabinovitch method gives an accurate parametric representation of the required density of states.

An alternative method of correcting the classical approximation, formulated by Haarhoff [43], is obtained by inverting the partition function without first taking the classical limit, i.e. by finding the inverse Laplace transform of the equation

$$Q_v(T) = \prod_{i=1}^{s} \frac{e^{-h\nu_i/2kT}}{1 - e^{-h\nu_i/kT}} = 2^s \prod_{i=1}^{s} \mathrm{cosech}(h\nu_i/2kT) \tag{87}$$

The cosech function are expanded as a power series in kT. Inversion of the Laplace transform then gives a series solution for the density of states, whose first term is the same as the Marcus-Rice correction, Eq. (85), and whose subsequent terms are corrections containing smaller powers of $E_v + E_z$. Ultimately the expansion of the partition function gives terms with negative powers of kT, which invert to delta functions and derivatives of delta functions, but truncating the series before this happens gives good smooth approximations to the density of states.

Expansion of the vibrational partition functions instead as geometric progressions in the exponentials $\exp(-h\nu_i/kT)$ gives the density of states as a sum of delta functions, each of which corresponds to a vibrational energy level. This is, of course, an exact representation of the density of states in the harmonic approximation, but it is not very convenient to work with analytically. However, a number of efficient computer algorithms have been developed for identifying and counting vibrational states. The case of harmonic oscillators is particularly simple because each oscillator has a ladder of equally spaced energy levels.

A very efficient method that divides energy into grains and performs the convolution of these ladders of energy levels for harmonic oscillators is the Beyer-Swinehart algorithm [44], which is stated formally in standard texts on the subject [25,29]. One inaccuracy of the method is that the difference between adjacent levels of the same oscillator is rounded to be an integer number of grains. As the vibrational quantum number increases this energy difference with the ground state of the oscillator is scaled by the vibrational quantum number, including the rounding error, so that the excited states are ultimately placed in the incorrect grains. This problem is generally smaller that the problems caused by assuming that the oscillators are harmonic, and can be alleviated by using small grains. It is perfectly feasible, and indeed common, to use grains as small as $1\ \mathrm{cm}^{-1}$, and then to lump grains together, if larger grains are required.

The magnification of the rounding error in the Beyer-Swinehart algorithm may also be removed by a method that calculates the energy levels of each oscillator before rounding them into energy grains, and then performing the convolution. Such a method was suggested by Stein and Rabinovitch [45], and can also be used for anharmonic oscillators, and other systems whose degrees of freedom are separable and whose energy levels are not equally spaced [46]. This algorithm is also stated and explained in [29].

In all the methods so far described it has been assumed that the vibrational modes of the molecule are separable from each other. At low energies, where the harmonic normal mode separation is valid this is a good approximation, but at higher energies the effect of anharmonicity not only affects the spacings between adjacent levels of the same oscillator, it also mixes vibrations and can have major effects on the energy levels. For example, one thinks immediately of Fermi resonances in this context. Of course it is exactly this mixing that the RRKM theory relies on to randomize the vibrational energy rapidly within the molecule and to maintain the microcanonical ensemble. Any method that is based on the convolution of independent stacks of vibrational levels ignores the effects of this type of mixing on the density of states.

For non-separable vibrational modes the only viable methods available are Monte Carlo methods, because of the high dimensionality of the sum to be evaluated. In this case it is much more efficient to calculate the sum of states, since it is more efficient to sample points within some high-dimensional volume than to identify those points within a thin shell. A method based on the random selection of vibrational quantum numbers has been proposed by Barker [47]. In this method the energy of a state depends on a set of values of each of s vibrational quantum numbers. Of course there must be some approximate formula for calculating the energy from this set of quantum numbers, for example a harmonic normal mode decomposition with corrections that are quadratic in the vibrational quantum number, like the Morse oscillator, but including cross terms. The Barker algorithm maximizes its efficiency by ensuring that each vibrational quantum number in turn is sampled from a range that will not exceed the energy being simulated, given the quantum numbers already assigned. A Monte Carlo integration for the classical sum of states based on a realistic potential energy surface has also been proposed, by Klippenstein and Marcus [48]. Although this method is classical, its results for the sum of states are in good agreement with the corresponding quantum mechanical system [49].

5.2.6. Quantum effects

Although the use of the correct energy levels for calculating the density of states is strictly a quantum correction of the classical RRKM theory, there are two other effects that are much more fundamental to the theory: quantum mechanical tunnelling and fluctuations. The first of these is dealt with in Chapter 2, and the second is the main subject of Chapter 3.

According to classical mechanics a trajectory with insufficient energy to overcome the energy barrier to reaction may not react. In quantum mechanics this is no longer the case, so that systems may still react, even though they do not have sufficient energy. This observation has several important consequences for RRKM theory. Firstly, it destroys the variational nature of the theory. So long as the theory overestimates $k(E,J)$ the rate coefficient may be optimized by minimization. However, we are now faced with situations where the theory predicts that $k(E,J) = 0$, but which still react. The RRKM rate coefficient is clearly no longer an upper bound on the true rate coefficient. Secondly, tunnelling does not necessarily follow the same path as the classical reaction coordinate. Indeed calculation of the tunnelling rate using the potential energy profile of the reaction coordinate may produce a severe underestimate [50]. The optimal path for tunnelling is a balance between the distance travelled and the barrier height. This balance may be improved by taking a path parallel to the reaction coordinate, but at the inner turning point of the vibrations, thus shortening the path, but at the expense of increasing the barrier height. This improvement was suggested by Marcus and Coltrin [51]. An entirely different method was suggested by George and Miller [53], in which a trajectory at its turning point switches into imaginary time and finds its own path through the classically forbidden region. A modification of this idea recognizes that mechanics in imaginary time is equivalent to mechanics in real time on an inverted potential [54]. These ideas are discussed further by Child [50]. The basic point, however, is that tunnelling does not follow the classical reaction path, and so the requirement of transition state theory that there be a unique reaction path, separable from all other motions, clearly breaks down.

For many unimolecular reactions tunnelling corrections are simply not necessary. Tunnelling is only likely to be important if the reaction coordinate involves transferring a small reduced mass over a barrier. Many reactions, such as associations do not have a barrier to surmount (apart from a small centrifugal barrier). However, when it is necessary, it is most convenient to calculate tunnelling corrections from a one-dimensional path, such as that of Marcus and Coltrin [51]. For a one-dimensional system the tunnelling rate must be calculated by solving the Schrödinger equation with the appropriate barrier potential. It is often convenient to fit this potential to a suitable parametric form for which tunnelling probabilities are known analytically, such as a parabolic barrier or an Eckart potential [52], which is of the general form

$$V(x) = \frac{V_0 e^{\alpha x}}{(1+e^{\alpha x})^2} + \frac{V_1 e^{\alpha x}}{1+e^{\alpha x}} \tag{88}$$

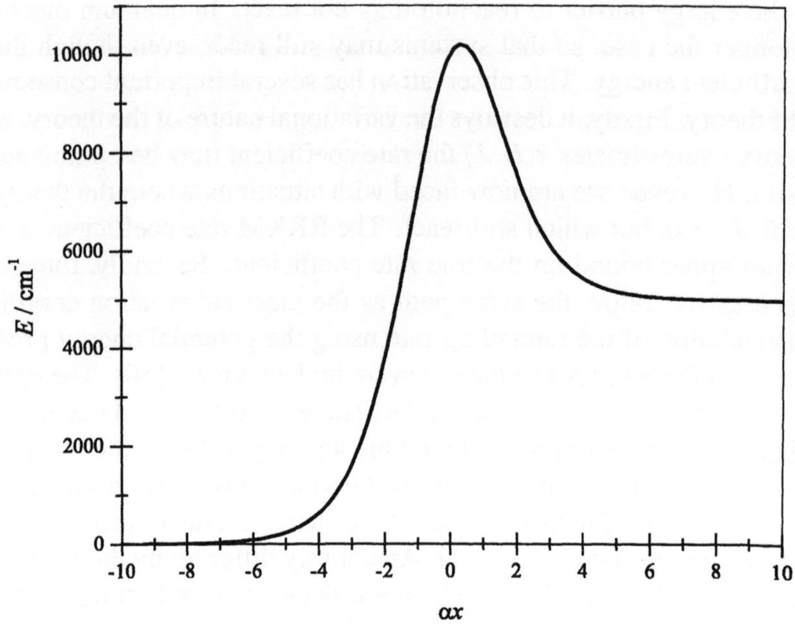

Figure 8. Example of an Eckart potential with an energy change of 5000 cm^{-1} and a barrier height of 10208 cm^{-1}.

This potential increases from 0 over a barrier of height $(V_0 + V_1)^2/4V_0$ and falls away again to a value V_1, as illustrated in Fig. 8.

The analytical solution for the transmission probability for this barrier is

$$T = \frac{\cosh\left(\beta(\sqrt{E} + \sqrt{E - V_1})\right) + \cosh\left(\beta(\sqrt{E} - \sqrt{E - V_1})\right)}{\cosh\left(\beta(\sqrt{E} + \sqrt{E - V_1})\right) + \cosh\left(\beta(\sqrt{8mV_0 - \pi^2/\beta^2})\right)} \tag{89}$$

where $\beta = 2\pi\sqrt{2m}/\hbar\alpha$. This type of one-dimensional approximation is very convenient, and is frequently used in applications, as discussed in Chapter 2.

Chapter 3 deals with an even more fundamental difficulty of RRKM theory. The theory expresses the microcanonical rate coefficient in terms of a sum and a density of states, and much of the effort in applying the theory goes into evaluating these quantities. However, the energies concerned are above the reaction threshold and so formally the molecule does not have any stationary states. The states above the threshold are resonance states, which are an extension of the sequence of stationary states into the continuum. Each resonance state is characterized by an energy and a width, which is related to its lifetime with respect to reaction by the uncertainty principle. One very important feature of the molecules for which these resonance lifetimes may be calculated quantum mechanically is that they exhibit fluctuations. That is, states of similar energies may have very different lifetimes, in complete contrast to the RRKM theory, which predicts that the rate coefficient increases with the sum of states, i.e. stepwise above the barrier. At present the full quantum theory can only be applied for relatively small systems, containing up to four atoms only, but this is an important area for future development.

6. STATISTICAL ADIABATIC CHANNEL MODEL

An alternative, and very attractive, approach to the microcanonical rate coefficient was proposed by Quack and Troe [60–64]. This theory, known as the Statistical Adiabatic Channel Model (SACM), is essentially a microcanonical version of vibrationally adiabatic transition state theory. The idea is that each individual state of reactants and products are linked by rovibrationally adiabatic channel potentials, which are obtained in principle by separating the reaction coordinate from the other molecular motions orthogonal to it, and solving the Schrödinger equation for their vibrational states at each point on the reaction coordinate. This approximation is similar to the Born-Oppenheimer approximation, except that it is less accurate, because the motion along the reaction coordinate, which is the frozen coordinate, is not generally much slower than the other motions being solved. In other words, all coupling between the reaction coordinate and the other vibrational motions is neglected, in complete contrast to the assumptions of RRKM theory.

Each reactant state correlates with some state of the products along the potential. Vibrations and rotations that are similar in the reactant and product (conserved modes), remain in the same quantum state throughout the channel, in the sense that their quantum numbers remain the same throughout. Other modes that change between reactants and products (transitional modes), are subject to correlation rules. Channels with the same angular momentum are not permitted to cross, similar to the non-crossing rule in diatomic molecules.

Channels may have barriers, even for systems where the potential energy surface does not have a barrier, such as a dissociation reaction. These barriers may arise for several reasons:

1. Centrifugal barriers, which arise because of the conservation of angular momentum along the channel.

2. Barriers that arise because some vibrational frequencies of the reactants are higher than those of the products, or may correlate with rotations in the products. For such modes the channels get closer together in energy, and so pass through a maximum.

3. Channels whose gradients have opposite signs and exhibit an avoided crossing develop a barrier

The last two of these concepts are illustrated in Fig. 9.

At a given energy E and angular momentum J the channels corresponding to angular momentum J are classified as either open or closed, depending on whether the energy available is sufficient to surmount the channel barrier. If the maximum energy of the adiabatic channel is E_{max} then the channel is open if $E > E_{\text{max}}$ and closed if $E < E_{\text{max}}$.

The assumption of strict adiabaticity is lifted by assuming that channels may be chosen statistically, according to a microcanonical ensemble. The rate coefficient is then expressed by a formula that is very similar to the RRKM result.

$$k(E,J) = \frac{W^{+}_{\text{SACM}}(E,J)}{h\rho(E,J)} \tag{90}$$

where $W^{+}_{\text{SACM}}(E,J)$ is the number of open channels with the required angular momentum. It is not necessary for each channel to be followed strictly, it is simply necessary for the adiabatic property to be obeyed statistically, i.e. that the count of the number of open channels be a good approximation.

There are several attractive features of the SACM that should be mentioned.

1. Because there is a unique one-to-one correlation of reactant states and product states over the reaction coordinate, the sum of the number of open channels may be made over the product states, rather than in the transition state region, as is necessary for RRKM theory.

2. Because the reactant states are selected from a microcanonical ensemble, the theory predicts which product states may be formed. The predicted

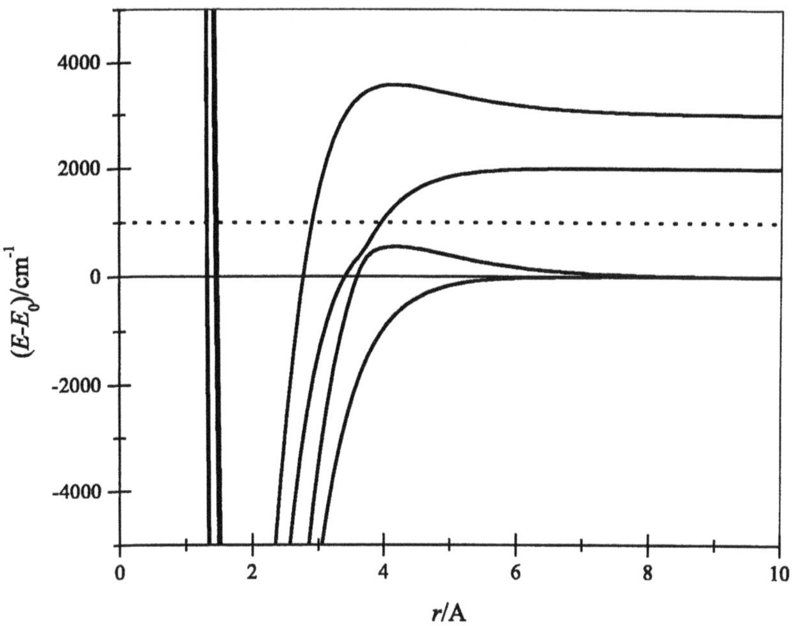

Figure 9. Schematic adiabatic channel potentials for a dissociation reaction, illustrating the development of barriers due to convergence of vibrational levels and the non-crossing rule. At the energy indicated, two of the channels are open and two are closed.

product state distribution may be quite good for reactions with loose transition states, where there is little mixing of channels on the product side. The product state distribution is not given by basic RRKM theory, although Klippenstein and Marcus have shown how RRKM theory may be extended to do this [65].

3. Channel potentials may be calculated for realistic or *ab initio* potential energy surfaces if a reaction coordinate can be defined.

4. Because the channels are essentially one-dimensional, it is easy to calculate tunnelling corrections, although, because the tunnelling path is not

the same as the reaction coordinate, simple corrections of this sort are likely to be underestimates, as discussed in section 5.2.6.

The main difficulty in applying the SACM is that an adiabatic channel must be calculated for each quantum state of the reactant. This level of detail is only possible for small molecules. For larger molecules the RRKM theory is more practical.

The basic assumptions of SACM appear to be entirely different from the RRKM theory. In SACM the adiabatic channels do not mix, whereas in RRKM theory the rate coefficient is calculated on the assumption of rapid redistribution of vibrational energy into the reaction coordinate. However, as stated above, the adiabatic property of SACM does not hold strictly, it is only necessary for the counting of channels to be correct. In this sense SACM can be thought of as a fully quantized version of RRKM theory, as stated in Chapter 2.

7. IMPROVED MODELS OF ENERGY TRANSFER

The previous sections have mainly focused on the reaction step of the Lindemann mechanism for a unimolecular reaction. However, to get the full pressure dependence of the thermal rate coefficient it is necessary to model the competition between deactivation and reaction accurately. The strong collision approximation, which leads to Eq. (21) goes a long way towards this end, but the assumption that the energy is completely Boltzmannized in every collision, and that almost every collision will de-energize an energized molecule, is not accurate. In order to reach the reaction threshold a molecule needs to gain tens of thousands of cm^{-1}, but it is well known that the typical energy transferred in a thermal collision with a small bath gas molecule is of the order of a few hundred cm^{-1} [66]. Energy transfer is treated in detail in Part 2 of this work. In consequence, it may take many collisions to energize or de-energize a molecule and a realistic model should recognize this.

The most important practical consequence of the failure of the strong collision approximation is that the fall-off curve is shifted to higher pressures, and broadened a little. There are two main approaches to dealing with this situation. It is possible to find parametric corrections for the weak collision effect, which will be introduced below. Alternatively the energy transfer may be treated as a random process with transitions between states occurring on collision. This latter type of approach is called the Master Equation, and is the main subject of Part 3 of this work.

7.1. Troe parameterizations

The parametric representation of unimolecular rate coefficients, both at the low pressure limit and in the fall-off region has been the result of a monumental work by Troe and his coworkers [67–71]. This work is tremendously useful for applications of unimolecular reactions to complex systems, for example to combustion or to atmospheric chemistry, as it permits accurate representation of the fall-off without having to to a time-consuming and difficult Master Equation calculation.

7.1.1. Low pressure limit

The low pressure limit (activation) rate coefficient is given in the strong collision approximation by

$$k_1(T) = \frac{Z}{Q_{vr}} \int_{E_0}^{\infty} \rho_{vr}(E_{vr}) \exp(-E/kT)\, dE_{vr} \tag{91}$$

Troe's starting point was to consider vibrations only and to approximate the density of states by its value at E_0, giving a zero-order approximation

$$k_1^{(0)}(T) = \frac{ZkT\beta_c}{Q_{vr}} \rho_{vr}(E_0) \exp(-E_0/kT) \tag{92}$$

in which an additional factor β_c, the collision efficiency, has been introduced in order to allow for the fact that collisions are not strong and that activation may take several collisions. The density of states is calculated using the Whitten-Rabinovitch method [42], of Eq. (86), which is an analytic semiclassical approximation.

This estimate is then improved by the inclusion of a series of correction factors, which were derived and developed by Troe [67,68].

$$k_1(T) = k_1^{(0)}(T) F_{\mathrm{anh}} F_{\mathrm{E}} F_{\mathrm{rot}} F_{\mathrm{rot,int}} F_{\mathrm{corr}} \tag{93}$$

Each of these correction factors takes into account one feature that is missing from the approximate formulation.

F_{anh} is a correction for the loss of vibrational energy due to the anharmonicity of the potential and due to the loss of vibrational modes in the dissociation reaction. Troe proposed the formula

$$F_{\mathrm{anh}} = \left(\frac{s-1}{s-3/2}\right)^m \tag{94}$$

where s is the total number of vibrational modes and m is the number of modes that are lost in the dissociation.

F_E is a correction for the fact that the density of vibrational states is an increasing function of energy above the threshold, whereas the approximate formula has assumed it to be constant. Approximating the density of states instead by the Whitten-Rabinovitch formula, Eq. (86), gives

$$F_E = \sum_{i=0}^{s-1} \frac{(s-1)!}{(s-i-1)!} \left(\frac{kT}{E_0 + aE_z} \right)^i \tag{95}$$

where a is the Whitten-Rabinovitch parameter [42].

F_{rot} is a correction that allows for the fact that only the vibrational density of states has been used, and the rotations also need to be included. The maximum correction factor is given by

$$F_{rot,max} = \frac{(s-1)!2^{s+1}}{(2s+1)!!\sqrt{\pi}} \left(\frac{E_0 + aE_z}{kT} \right)^{3/2} \tag{96}$$

where the double factorial symbol designates the product of all odd numbers up to $2s+1$.

This correction factor is reduced to allow for the appearance of centrifugal barriers [29,72].

$$F_{rot} = F_{rot,max} \frac{I^\dagger/I}{I^\dagger/I - 1 + F_{rot,max}} \tag{97}$$

The ratio of the moment of inertia of the transition state to that at the equilibrium geometry depends strongly on the shape of the potential curve.

$F_{rot,int}$ takes care of any internal rotors. In general, this correction can be quite complicated because the rotation may be hindered. However, in the simplest case of a single internal free rotor the correction factor is

$$F_{rot,int} = \frac{(s-1)!2^{s}}{(2s-1)!!\sqrt{\pi}} \left(\frac{E_0 + aE_z}{kT} \right)^{1/2} \tag{98}$$

More general cases are considered in standard texts on the subject [73,29]

F_{corr} is a correction factor that allows for the possibility that the other factors considered so far may not be independent of one another. This factor is generally close to 1, but it may be different if barriers arise as a result of changes in the frequencies of transitional modes.

The collision efficiency parameter β_c is related to the mean energy change in a collision ΔE by the formula [67,68]

$$\frac{\beta_c}{1-\sqrt{\beta_c}} = \frac{-\Delta E}{F_E kT} \tag{99}$$

This parameter is generally used to fit experimental data, and the values obtained in this way are physically reasonable.

7.1.2. Parameterization of the fall-off

Troe proposed a similar approach to the calculation of the fall-off curve. In this case the zero-order approximation is the Lindemann-Hinshelwood model, formulated with the correct high and low-pressure limiting rate coefficients:

$$k_{\text{uni}}^{(0)} = \frac{k_\infty k_1 [\text{M}]}{k_\infty + k_1 [\text{M}]} \tag{100}$$

The shape of the fall-off must then be corrected, both for the energy-dependence of the microcanonical rate coefficient in the strong collision approximation, and for the fact that the collisions are not, in fact, strong.

$$k_{\text{uni}} = k_{\text{uni}}^{(0)} F_{\text{SC}} F_{\text{WC}} \tag{101}$$

The strong collision correction factor F_{SC} is a function of two further parameters that arise in Kassel theory (see Section 3). These are the number of effective oscillators, S, and $B = E_0/kT$, a measure of the relative magnitudes of the threshold energy and thermal energy. Troe used an energy criterion to obtain the number of effective oscillators,

$$S_K = 1 + E_v^\dagger / kT \tag{102}$$

where E_v is the mean vibrational energy of the molecule, which can be calculated by elementary statistical mechanics.

The other parameter was again modified to take account of the effective number of oscillators and the zero point energy, using the Whitten-Rabinovitch formula

$$B_K = \frac{S_K - 1}{s-1} \frac{(E_0 + aE_z)}{kT} \tag{103}$$

Troe parameterized the pressure-dependence of F_{SC} empirically in terms of a central value and a width parameter. The central value is given by

$$\log F_{SC,cent} = \frac{-(1.06 \log S_K)^{2.2}}{1 + C_1 S_K^{C_2}} \tag{104}$$

where

$$C_1 = 0.10 \exp(2.5/B_T - 0.22 B_T - 6 \times 10^{-10} B_T^6) \tag{105}$$

$$C_2 = 1.9 + 4.6 \times 10^{-5} B_T^{2.8} \tag{106}$$

$$B_T = \left(\frac{S_K - 1}{s - 1}\right)^{0.6} B_K \tag{107}$$

Similarly, the central value for the weak collisions correction is

$$\log F_{WC,cent} = 0.14 \log \beta_c \tag{108}$$

although a more detailed expression may be necessary at high temperatures [71].

It is convenient to combine the pressure dependence of the two correction factors. Defining $F_{cent} = F_{SC,cent} F_{SC,cent}$, the pressure dependence of the combined correction F is given by [29]

$$\frac{\log F}{\log F_{cent}} = \frac{1}{1 + [\kappa/(N - 0.14\kappa]^2} \tag{109}$$

where

$$\kappa = \log(k_1[M]/k_\infty) + c \tag{110}$$

$$c = -0.4 - 0.67 \log F_{cent} \tag{111}$$

$$N = 0.75 - 1.27 \log F_{cent} \tag{112}$$

Although these equations are empirical, they do give a good representation of experimental fall-off curves, and are frequently used to analyse and represent data.

7.2. Master equation

The alternative to using a parametric method is to formulate a more precise description of the energy transfer. The basic physical picture underlying the model is that a molecule travels inertially between collisions, which take place with some pseudo-first order frequency $\omega = Z[\mathrm{M}]$. The duration of the collision is assumed to be much smaller than the waiting time between collisions. On collision the molecule undergoes a transition from one state to another. The molecular state is usually designated by its energy, although the angular momentum label may also be used. Suppose that the probability density of the energy E after a collision given that the energy before the collision was E' is denoted $P(E|E')$. The population density of level E, $p(E,t)$, then obeys a rate equation of the form

$$\frac{\partial p}{\partial t} = \omega \int_0^\infty P(E|E')p(E',t)\,dE' - \omega p(E,t) k(E) p(E,t) \tag{113}$$

The first term represents the rate of transitions to E from all other energies on collision. The second term is the rate at which collisions remove density from energy E and the third term is the rate at which density at energy E is removed by reaction, $k(E)$ being the microcanonical rate coefficient.

This Master Equation is turned into a finite set of coupled linear differential equations by truncating the energy space at some high value that is only visited very infrequently, and graining the energy.

$$\frac{d\mathbf{p}}{dt} = \omega(\mathbf{P} - \mathbf{I})\mathbf{p} - \mathbf{K}\mathbf{p} \tag{114}$$

Then numerical methods of matrix diagonalization are used to find the eigenvalues of the matrix operator $\omega(\mathbf{P} - \mathbf{I}) - \mathbf{K}$, which are the time constants that determine both the chemical kinetics and the energy relaxation. Part three of this work deals in detail with the formulation of the Master Equation for a number of different systems, for example termolecular association reactions and reversible reactions. It then deals with methods for finding the time constants and simulating the kinetics. The Master Equation is the method of choice at present for modelling the competition between energy transfer and reaction.

REFERENCES

[1] J. Perrin, Ann. Phys. 11 (1919) 1.
[2] I. Langmuir, J. Amer. Chem. Soc. 42 (1920) 2190.
[3] R.C. Dunbar and T.B. McMahon, Science 279 (1998) 194.
[4] F.A. Lindemann, Trans. Faraday Soc. 17 (1922) 598.
[5] J.A. Christiansen, Thesis, University of Copenhagen, 1921.
[6] C.T. Genaux, F. Kern and W.D. Walters, J. Amer. Chem. Soc. 75 (1953) 6196.

[7] J.N. Butler and R.B. Ogawa, J. Amer. Chem. Soc. 85 (1963) 3346.
[8] M. Abramowitz and I.A. Stegun, Handbook of Mathematical Functions, Dover, New York, 1965.
[9] C.N. Hinshelwood, Proc. Roy. Soc. A113 (1927) 230.
[10] O.K. Rice and H.C. Ramsperger, J. Amer. Chem. Soc. 49 (1927) 1617.
[11] O.K. Rice and H.C. Ramsperger, J. Amer. Chem. Soc. 50 (1928) 617.
[12] L.S. Kassel, J. Phys. Chem. 32 (1928) 225.
[13] L.S. Kassel, J. Phys. Chem. 32 (1928) 1065.
[14] D.W. Placzek, B.S. Rabinovitch and G.Z. Whitten, J. Chem. Phys. 43 (1965) 4071.
[15] S.W. Benson, Thermochemical Kinetics, Wiley, New York, 1968.
[16] Ber. Bunsenges. Phys. Chem. 71 (1967) 937.
[17] G.B. Skinner and B.S. Rabinovitch, J. Phys. Chem. 76 (1972) 2418.
[18] H.W. Schranz, S. Nordholm and N.D. Hamer, Int. J. Chem. Kinet. 14 (1982) 543.
[19] J.C. Giddings and H. Eyring, J. Chem. Phys. 22 (1954) 538.
[20] R.A. Marcus and O.K. Rice, J. Phys. and Colloid Chem. 55 (1951) 894.
[21] R.A. Marcus, J. Chem. Phys. 20 (1952) 359.
[22] G.N. Wieder and R.A. Marcus, J. Chem. Phys. 37 (1962) 1835.
[23] R.A. Marcus, J. Chem. Phys. 43 (1965) 2658.
[24] Chem. Phys. 10 (1975) 403.
[25] R.G. Gilbert and S.C. Smith, Theory of Unimolecular and Recombination Reactions, Blackwell, Oxford, 1990.
[26] D.A. McQuarrie, Statistical Mechanics, Harper and Row, New York, 1976.
[27] W. Forst, Theory of Unimolecular Reactions, Academic Press, New York, 1973.
[28] N.J.B. Green, Quantum Mechanics 1: Foundations, Oxford University Press, Oxford, 1997.
[29] K.A. Holbrook, M.J. Pilling and S.H. Robertson, Unimolecular Reactions, Wiley, Chichester, 1996.
[30] J.O. Hirschfelder and E. Wigner, J. Chem. Phys. 7 (1939) 616.
[31] V. Gold, Trans. Faraday Soc. 60 (1964) 739.
[32] E.W. Schlag and G.L. Haller, J. Chem. Phys. 42 (1965) 584.
[33] C.S. Elliott and H.M. Frey, Trans. Faraday Soc. 664 (1968) 2352.
[34] D.M. Bishop and K.J. Laidler, J. Chem. Phys. 42 (1965) 1688.
[35] P. Pechukas in Dyamics of Molecular Collisions B, Ed. W.H. Miller et al., Plenum, New York, 1976.
[36] J.C. Keck, J. Chem. Phys. 53 (1970) 2041.
[37] B.C. Garrett and D.G. Truhlar, J. Chem. Phys. 70 (1979) 1593.
[38] D.G. Truhlar and B.C. Garrett, Acc. Chem. Res. 13 (1980) 440.
[39] D.M. Wardlaw and R.A. Marcus, J. Chem. Phys. 83 (1985) 3462.
[40] D.M. Wardlaw and R.A. Marcus, J. Phys. Chem. 90 (1986) 5383.
[41] S.H. Robertson and D.M. Wardlaw, Chem. Phys. Lett. 199 (1992) 391.
[42] G.Z. Whitten and B.S. Rabinovitch, J. Chem. Phys. 38 (1963) 2466.
[43] P.C. Haarhoff, Molec. Phys. 6 (1963) 337.
[44] T. Beyer and D.F. Swinehart, Comm. Assoc. Comput. Mach. 16 (1973) 379.
[45] S.E. Stein and B.S. Rabinovitch, J. Chem. Phys. 58 (1973) 2438.
[46] S.E. Stein and B.S. Rabinovitch, J. Chem. Phys. 60 (1974) 908.
[47] J.R. Barker, J. Phys. Chem. 91 (1987) 3849.

[48] S.J. Klippenstein and R.A. Marcus, J. Chem. Phys. 87 (1987) 3410.
[49] S.J. Klippenstein and R.A. Marcus, J. Phys. Chem. 92 (1988) 5412.
[50] M.S. Child in Modern Gas Kinetics, Ed. M.J. Pilling and I.W.M Smith, Blackwell, Oxford, 1987, p. 18.
[51] R.A. Marcus and M.E. Coltrin, J. Chem. Phys. 67 (1977) 5668.
[52] C. Eckart, Phys. Rev. 35 (1930) 1303.
[53] T.E. George and W.H. Miller, J. Chem. Phys. 57 (1972) 2458.
[54] S. Chapmen, R.C. Garrett and W.H. Miller, J. Chem. Phys. 63 (1975) 2710.
[55] N.B. Slater, Proc. Cambridge Phil. Soc. 35 (1959) 56.
[56] N.B. Slater, Theory of Unimolecular Reactions, Cornell University Press, Ithaca, 1959.
[57] E.B. Wilson Jr, J.C. Decius and P.C. Cross, Molecular Vibrations, Dover, New York, 1980.
[58] M. Kac, Amer. J. Math. 65 (1943) 609.
[59] M. Kac, Proc. London Math. Soc. 50 (1949) 390.
[60] M. Quack and J. Troe, Ber. Bunsenges. Phys. Chem. 78 (1974) 240.
[61] M. Quack and J. Troe, Ber. Bunsenges. Phys. Chem. 79 (1975) 170.
[62] M. Quack and J. Troe, Ber. Bunsenges. Phys. Chem. 79 (1975) 469.
[63] M. Quack and J. Troe, Ber. Bunsenges. Phys. Chem. 81 (1977) 329.
[64] M. Quack and J. Troe, Int. Rev. Phys. Chem. 1 (1977) 97.
[65] S.J. Klippenstein and R.A. Marcus, J. Chem. Phys. 91 (1989) 2280.
[66] H. Hippler, J. Troe and H.J. Wendelken, J. CHem. Phys. 78 (1983) 6709.
[67] J. Troe, J. Chem. Phys. 66 (1977) 4745.
[68] J. Troe, J. Chem. Phys. 66 (1977) 4758.
[69] J. Troe, J. Phys. Chem. 83 (1979) 118.
[70] J. Troe, Ber. Bunsenges. Phys. Chem. 78 (1974) 478.
[71] R.G. Gilbert, K. Luther and J. Troe, Ber. Bunsenges. Phys. Chem. 87 (1983) 169.
[72] E.V. Waage and B.S. Rabinovitch, Chem. Rev. 70 (1970) 37.
[73] I.W.M. Smith in Modern Gas Kinetics, Ed. M.J. Pilling and I.W.M Smith, Blackwell, Oxford, 1987, p. 99.

Chapter 2

RRKM Theory and Its Implementation

Stephen J. Klippenstein

Combustion Research Facility, Sandia National Laboratories, Livermore, California, USA, 94551-0969

2.1. BACKGROUND

The development of RRKM theory (named after Rice, Ramsperger, Kassel and Marcus) [1-4] is one of the major triumphs of theoretical chemistry. As discussed in Ch. 1, RRKM theory was originally derived to treat the temperature and pressure dependence of unimolecular dissociation (and/or isomerization) reactions. It coupled a quantum transition state theory (TST) treatment of the collisionless dissociation process with a strong collider model for the energy transfer process. The resulting theory proved to be very effective at modeling a wide variety of experimental results [5, 6]. Ultimately, however, it became clear that the strong collider assumption is overly simplistic and must be replaced with a more detailed model for the energy dependence of the collision-induced energy transfer process [7, 8]. The development of such explicit energy transfer models, and their incorporation within master equation simulations, is the subject of parts 2 and 3 of this monograph, respectively.

With this replacement of the strong collider assumption now commonplace, the term RRKM theory has become largely synonymous with quantum TST for unimolecular reactions, and we use this terminology here. The foundations of RRKM theory have been tested in depth with a wide variety of inventive theoretical and experimental studies [9]. While these tests have occasionally indicated certain limitations in its applicability, for example to timescales of a picosecond or longer, the primary conclusion remains that RRKM theory is quantitatively valid for the vast majority of conditions of importance to chemical kinetics. The $H + O_2 \leftrightarrow HO_2 \leftrightarrow OH + O$ reaction is an example of an important reaction where deviations from RRKM predictions are significant [10, 11]. The foundations of RRKM theory and TST have been aptly reviewed in various places [7, 9, 12-15]. Thus, the present chapter begins with only a brief review of the dynamical foundations of RRKM theory; primarily as a means for motivating various aspects of some of the recent implementations.

RRKM theory has gradually become one of the most effective and commonly used tools for chemical kineticists. Indeed, its applications are so widespread as to be effectively innumerable. In recent years, these applications have evolved from primarily empirical models for interpreting, interpolating, and extrapolating experimental data, to making truly *a priori* predictions. This transformation has largely been the result of the continuing advances in quantum chemistry. With this transformation, there has been considerable effort devoted to improving the ease and accuracy of *a priori* implementations of RRKM theory. These improvements include more flexible descriptions of the transition state dividing surface, more quantitative incorporations of anharmonic effects, and, perhaps most importantly, the development of procedures for coupling together the quantum chemical and partition function evaluations.

The focus of this chapter is a review of the methodologies employed in *a priori* implementations of RRKM theory for the collisionless dissociation/ isomerization steps in gas-phase unimolecular reactions. Special attention will be paid to recent developments, particularly those that have proven their utility through substantive applications. With microscopic reversibility, RRKM treatments of the dissociation process are directly applicable to the reverse bimolecular associations. Furthermore, some of the more interesting illustrations are for bimolecular reactions and so we do not limit our discussion of RRKM theory to unimolecular reactions. However, one should bear in mind that TST was originally derived for bimolecular reactions and the specific term RRKM theory is really only applicable to the unimolecular direction.

The presence or absence of a well-defined saddlepoint is an important distinguishing feature in implementations of RRKM theory. With a well-defined saddlepoint variational effects tend to be relatively minor, while an adequate treatment of tunneling can be of key importance. In contrast, in the absence of a saddlepoint, tunneling can generally be ignored, while variational effects are quite significant. Furthermore, there tend to be a number of large amplitude motions suggesting the importance of anharmonic effects. Thus, implementations for these two classes of reactions are discussed separately below. In these discussions we consider solely the evaluation of the transition state partition functions. The accurate *a priori* evaluation of the reactant partition functions tends to be comparatively straightforward, although significant anharmonic effects may arise in some instances. Thus, we also provide a review of procedures for evaluating such anharmonic effects on the reactant partition functions.

2.2. DERIVATION OF RRKM THEORY

2.2.1. Classical RRKM theory

It is instructive to begin with a derivation of RRKM theory from the classical dynamical expression for the rate coefficient. The latter rate coefficient, $k_{\text{classical}}(t)$, is generally time dependent, and may be expressed as

$$k_{\text{classical}}(t) = \frac{\int d\mathbf{q}d\mathbf{p}P_{\text{init}}(\mathbf{q},\mathbf{p})\delta\left(T(\mathbf{q},\mathbf{p})-t\right)}{\int d\mathbf{q}d\mathbf{p}P_{\text{init}}(\mathbf{q},\mathbf{p})\Theta\left(T(\mathbf{q},\mathbf{p})-t\right)}, \tag{2.1}$$

where $P_{\text{init}}(\boldsymbol{q},\boldsymbol{p})$ is the initial probability distribution in phase space, $T(\boldsymbol{q},\boldsymbol{p})$ is the time at which the trajectory initiated at the phase space point $(\boldsymbol{q},\boldsymbol{p})$ at $t=0$ transforms from reactants into products, and Θ and δ are the Heaviside step function and Dirac delta function, respectively.

If the classical dynamics is ergodic, with a chaotic sampling of all phase space on a timescale that is short compared to the dissociation timescale, then the rate coefficient is independent of time. Furthermore, the rate coefficient may then be expressed as an average, over the statistical distribution $P_{\text{stat}}(\boldsymbol{q},\boldsymbol{p})$, of the flux through an arbitrary dividing surface ($S=0$) separating reactants ($S<0$) from products ($S>0$)

$$k_{\text{ergodic}} = \int d\mathbf{q}d\mathbf{p}P_{\text{stat}}(\mathbf{q},\mathbf{p})\delta\left(S\right)\dot{S}\chi_{R}(\mathbf{q},\mathbf{p}). \tag{2.2}$$

The dividing surface may depend on both coordinates **q** and momenta **p** [16]. The quantity $\dot{S}$ is the velocity through the dividing surface, and the reactivity function, χ_R, is defined to be unity if a trajectory initiated at the given phase space point on the dividing surface proceeds directly on to products and zero otherwise [17].

The key further assumption of RRKM theory is that there is no recrossing of the dividing surface. In other words, it is assumed that all trajectories passing through the transition state in the direction of products continue directly on to products. With this assumption, the reactivity function is approximated by a step function in the velocity through the dividing surface, yielding

$$k_{\text{RRKM}} = \int d\mathbf{q}d\mathbf{p}P_{\text{stat}}(\mathbf{q},\mathbf{p})\delta\left(S\right)\dot{S}\Theta(\dot{S}). \tag{2.3}$$

It is important to note that this assumption yields an RRKM rate coefficient, k_{RRKM}, that is an upper bound to the ergodic rate coefficient, k_{ergodic}, since every reactive trajectory (with $\chi_R=1$) necessarily has a positive velocity through the dividing surface. Thus, RRKM theory may be implemented in a variational manner, with the best approximation to k_{ergodic} obtained from the dividing surface S that provides the smallest k_{RRKM}.

It is also worth emphasizing that RRKM theory is a local theory, requiring and using information about the potential energy surface only in the region of the transition state dividing surface. In contrast, dynamical theories require potential energy surface information for the full range of configuration space sampled by the dynamics. Importantly, this locality property has facilitated the coupling of quantum chemical simulations with RRKM theory, and is largely responsible for its development into such an invaluable tool for practical kinetic theory.

In a collisionless environment, the total energy E and the total angular momentum J are conserved quantities. The appropriate statistical distribution is then

$$P_{\mathrm{stat}}(\mathbf{q},\mathbf{p})=\frac{\delta\left(E-H(\mathbf{q},\mathbf{p})\right)\delta\left(J-\hat{J}(\mathbf{q},\mathbf{p})\right)}{\int d\mathbf{q}d\mathbf{p}\,\delta\left(E-H(\mathbf{q},\mathbf{p})\right)\delta\left(J-\hat{J}(\mathbf{q},\mathbf{p})\right)\Theta(-S)}, \tag{2.4}$$

where H is the Hamiltonian and $\hat{J}$ is the dynamical expression for the total angular momentum. The expression in the denominator is proportional to the classical density of states, $\rho(E,J)$, for the reactant region of phase space. Thus, the E and J resolved classical RRKM rate coefficient for a unimolecular reaction may be written as [18-20]

$$k_{\mathrm{RRKM}}^{\mathrm{classical}}(E,J)=\frac{\hbar}{\rho(E,J)}\int\frac{d\mathbf{q}d\mathbf{p}}{h^n}\delta\left(E-H(\mathbf{q},\mathbf{p})\right)\delta\left(J-\hat{J}(\mathbf{q},\mathbf{p})\right)\delta(S)\dot{S}\Theta(\dot{S}), \tag{2.5}$$

where h is Planck's constant and n is the dimension of configuration space. Eq. (2.5) provides the key starting point for variational RRKM evaluations.

The dividing surface is commonly represented in terms of a fixed value for some particular reaction coordinate R. When the kinetic energy and total angular momentum are separable in this coordinate the rate coefficient can be written in its standard RRKM form [1-6]

$$k_{\mathrm{RRKM}}(E,J) = \frac{N^{\dagger}(E,J)}{h\rho(E,J)}, \tag{2.6}$$

where $N^{\dagger}(E,J)$=min$\{N(E,J,R);R\}$=$N(E,J,R^{\dagger})$ is the number of states with energy less than E and with a total angular momentum of J, for motion on the transition state dividing surface specified by R=$R^{\dagger}$. The classical expression for $N(E,J,R)$ is given by

$$N(E,J,R) = \int \frac{d\mathbf{q}'d\mathbf{p}'}{h^{n-1}} \Theta\left(E - H(\mathbf{q}',R,\mathbf{p}',p_R = 0)\right)\delta\left(J - \hat{J}(\mathbf{q}',R,\mathbf{p}',p_R = 0)\right), \tag{2.7}$$

where p_R is the momentum conjugate to R, and $(\boldsymbol{q}) = (\boldsymbol{q}',R)$ while $(\boldsymbol{p}) = (\boldsymbol{p}',p_R)$.

2.2.2. Quantum RRKM theory

The classical RRKM expression provides rate coefficients that are generally in error by orders of magnitude. The simple replacement of the classical expressions for the transition state number of states and the reactant density of states with their quantum mechanical counterparts

$$N(E,J,R) = \sum_{i=1}^{\infty} \Theta\left(E - E_i(R)\right), \tag{2.8}$$

$$\rho(E,J) = \sum_{i=1}^{\infty} \delta\left(E - E_i\right). \tag{2.9}$$

yields quantum RRKM rate coefficients that are generally in good agreement with experiment. The sums in Eqs. (2.8) and (2.9) range over the quantized energy levels for motion on the dividing surface and in the reactant phase space, respectively, and are for states with a total angular momentum of J. Note that vibrationally adiabatic models [21], employing step functions in the reaction probabilities at the energies of the adiabatic channel maxima, are in essence identical to variational RRKM theory, at least for rate constant predictions. The reaction coordinate dependent adiabatic channel energies $E_i(R)$ are determined from diagonalizations of the Hamiltonian for the modes orthogonal to the reaction coordinate.

The quantization of the transition state number of states (Eq. (2.8)) can be justified for a separable reaction coordinate or for a vibrationally adiabatic model [21, 22]. Furthermore, in these cases, one may also include the effect of quantum mechanical tunneling through the replacement of the step functions in Eq. (2.8) with one-dimensional tunneling probabilities $P_{tun}(E)$

$$N(E,J,R)=\sum_{i=1}^{\infty}P_{\rm tun}\left(E-E_i\right). \qquad (2.10)$$

The quantized results of Eqs. (2.8) and (2.10) negate, at least in principle, the variational property of classical RRKM theory, since both the one-dimensional tunneling probabilities and the step function may exceed the true reactivities. In practice, however, variational minimizations still prove useful.

Various attempts have been made to derive fully quantized versions of TST appropriate to the nonseparable and nonadiabatic case [23-30]. These efforts are invariably plagued by the nonlocal nature of tunneling. An accurate treatment of tunneling necessitates the reintroduction of dynamics, at least to cover the generally quite extensive range of the tunneling motion, thereby removing the great value of a local theory. Thus, practical quantum TST theories have resorted to some approximate one-dimensional tunneling path that effectively accounts for the optimal tunneling path in the full multidimensional phase space [15, 31, 32]. In contrast with the difficulties in deriving and applying non-separable quantum TST, the classical expression provided in Eq. (2.5) has been effectively used in various treatments of non-separable reaction coordinates [33-35]. The coupling of classical treatments for a limited set of low-frequency modes, including the reaction coordinate, with quantum treatments for the remaining high frequency modes provides an effective treatment when tunneling is unimportant, as with barrierless reactions [36, 37].

2.2.3. Limitations

At this point it is worthwhile to review the possible failures of RRKM theory [9, 14] within a classical framework. First, the dynamics in some regions of phase space may not be ergodic. In this instance, which has been termed "intrinsic" non-RRKM behaviour [38], the use of the statistical distribution in Eq. (2.2) is inappropriate. In the extreme case of two disconnected regions of space, with one region nonreactive, the lifetime distribution is still random with an exponential decay of population to a non-zero value. However, the averaging of the flux must then be restricted to the reactive part of the phase space, and the rate coefficient is then increased by a factor equal to the reciprocal of the proportion of the phase space that is reactive.

Alternatively, the initial preparation process (e.g., collisional excitation, photoexcitation, or chemical activation) may not provide a random initial distribution. Then if the initial decay rate is rapid, i.e., faster than the rate of energy randomization, then its value may depend on the details of the initial distribution. This "apparent" non-RRKM behaviour often limits the applicability of RRKM theory to reactions occuring on a picosecond or longer timescale. This limitation arises most commonly in photodissociation experiments where the initial distribution is far from random, and the excitation can be to high energies where the dissociation is rapid. It can also be important when considering chemically activated reactants producing highly exothermic products. In contrast, collisional excitation is generally considered to serve as an effective randomizer, reducing the extent of both apparent and intrinsic non-RRKM effects. Comparisons of trajectory simulations with RRKM theory often indicate significant non-RRKM behaviour [9]. However, it is worth noting that due to computational limitations such simulations are typically restricted to timescales of about 1 ps or less, and do not generally include collisional randomization effects. Thus, such comparisons may be somewhat biased in their indication of non-RRKM effects.

A third possible failure of RRKM theory involves the presence of trajectories that recross the transition state dividing surface. In variational RRKM theory the effect of such recrossing is reduced via the consideration of multiple dividing surfaces. The larger the set of dividing surfaces considered, the better the estimate. However, each additional dividing surface requires additional potential energy surface information, thereby effectively removing, at least to some extent, the great advantage of the locality of RRKM theory. Thus, practical implementations of RRKM theory have generally considered limited families of dividing surfaces. An in depth understanding of the physical factors determining the structure of the optimal transition state dividing surface is thus of some value in maintaining the local nature of the potential energy surface requirements.

2.2.4. Canonical and microcanonical RRKM theory

Applications of RRKM theory often focus on the high pressure limit of the dissociation rate coefficient. The presence of multiple collisions prior to reaction generally maintains a Boltzmann population distribution. Correspondingly, the thermal rate coefficient is then expressed as a Boltzmann average over the energy and angular momentum resolved rate coefficient, which reduces to

$$k^{\infty}_{\mathrm{RRKM}}(T) = \frac{\int dEdJ \exp(-\beta E) N^{\dagger}(E,J)}{hQ(T)}, \tag{2.11}$$

where $Q(T)$ is the canonical partition function for the reactants

$$Q(T)=\sum_{i=1}^{\infty}\exp(-\beta E_i), \tag{2.12}$$

with $\beta=1/(k_BT)$. The Boltzmann average in Eq. (2.11) can also be viewed as a Laplace transform.

With this expression the minimization of the number of states for each E and J is performed prior to the integration over these variables. Reversing the order of these operations yields the standard canonical RRKM result

$$\begin{aligned} k_{\mathrm{RRKM}}^{\infty,\mathrm{canonical}}(T) &= \frac{\min\left\{\int dE\exp(-\beta E)N(E,R);R\right\}}{hQ(T)} \\ &= \frac{k_BT}{h}\frac{Q^{\dagger}(T)}{Q(T)} \end{aligned}, \tag{2.13}$$

where $Q^{\dagger}(T)=\min\{Q(T,R);R\}$, $Q(T,R)$ is the canonical partition function for motion on the transition state dividing surface specified by R, and $N(E,R)$ is the integral over J of $N(E,J,R)$.

The minimization of the canonical transition state partition function as in Eq. (2.13) is generally termed canonical variational RRKM theory. This approach provides an upper bound to the more proper E/J resolved minimization, but is still commonly employed since it simplifies both the numerical evaluation and the overall physical description. It typically provides a rate coefficient that is only 10 to 20% greater than the E/J resolved result of Eq. (2.11).

Occasionally, the order of integration and minimization is reversed for only the total angular momentum. This reversal provides a microcanonical variational RRKM result that is intermediate between the canonical and E/J resolved results

$$k_{\mathrm{RRKM}}^{\infty,\mu}(T)=\frac{\int dE\exp(-\beta E)\min\{N(E,R);R\}}{hQ(T)}. \tag{2.14}$$

2.3. REACTIONS WITH BARRIERS

2.3.1. Conventional RRKM theory with Eckart tunneling

For reactions with a well defined barrier the saddle point on the potential energy surface provides a good first approximation to the minimum energy configuration on the transition state dividing surface. Thus, in the quantum chemical literature the saddle point is often labelled the transition state. A simple, and quite useful, "conventional" RRKM estimate for the rate coefficient is obtained from a normal mode analysis at the saddle point. The quadratic expansion of the potential energy surface at the saddle point yields standard harmonic-oscillator vibrational frequencies (ν_i, $i=1,...,n$), but with that for the reactive mode, ν_n, having an imaginary frequency. The Beyer-Swinehart algorithm [39] provides a simple and efficient procedure for evaluating the harmonic-oscillator vibrational contribution to the number of available states at the transition state, $N^{\dagger}_{\mathrm{vib}}(E)$:

$$N^{\dagger}_{\mathrm{vib}}(E) = \sum_{v_i=1}^{\infty} \Theta\left(E - E_{\mathrm{vib}}(v_1, v_2, ... v_{n-1})\right). \tag{2.15}$$

Conventional RRKM theory estimates for $N^{\dagger}(E,J)$ are then obtained via a convolution with the rigid rotor rotational energy levels, $E^{\dagger}_{rot}(J,K)$, for the molecular structure at the stationary point:

$$N^{\dagger}(E,J) = \sum_{K} (2J+1) N^{\dagger}_{vib}\left(E - E^{\dagger}_{rot}(J,K)\right). \tag{2.16}$$

At the canonical level, the convolution theorem of Laplace transforms implies that the convolution over vibration and rotation reduces to a simple product:

$$Q^{\dagger}(T) = Q^{\dagger}_{vib}(T) Q^{\dagger}_{rot}(T). \tag{2.17}$$

Analytic expressions are available for both the quantum harmonic vibrator partition function $Q^{\dagger}_{\mathrm{vib}}(T)$ and the classical rigid rotor partition function $Q^{\dagger}_{\mathrm{rot}}(T)$ [40]; further simplifying the analysis.

While conventional RRKM theory is often quite accurate it fails in the low temperature limit due to tunneling through the potential barrier. The temperature at which such tunneling corrections become insignificant is highest for the transfer of low mass particles such as an H atom. For other motions with a greater reduced mass the tunneling corrections are generally insignificant at room temperature and higher. For separable reaction coordinates the tunneling probability is easily evaluated and is often expressed in terms of a barrier penetration integral, θ:

$$P_{tun}(E) = \left(1 + \exp(2\theta)\right)^{-1}. \tag{2.18}$$

For the parabolic barrier arising from the quadratic expansion about the saddle point, the barrier penetration integral is given by

$$\theta = \frac{\pi\left(V^{\dagger} - E\right)}{h\left|\nu_n\right|}, \qquad (2.19)$$

where $V^{\dagger}$ is the zero-point corrected potential at the saddle point and ν_n is the imaginary frequency for motion through the potential barrier. Thermal averaging yields a canonical parabolic barrier tunneling correction factor, Γ, for the transition state partition function given by [41]

$$\Gamma = \frac{|u|/2}{\sin\left(|u|/2\right)}, \qquad (2.20)$$

where $u \equiv h\nu_n/k_BT$. Unfortunately, this expression is of limited value since it diverges at low temperatures (i.e., $k_BT < h\nu_n/2\pi$), which is precisely where tunneling corrections are of most importance.

This divergence arises because the parabolic barrier allows for tunneling from infinitely low energies. Improved estimates can be obtained with similar computational effort via the consideration of the deviation from parabolic behaviour for the reactive mode potential at large displacements. The Eckart potential has the correct asymptotic properties of the potential for a direct reaction, and its parameters can be specified in terms of the curvature of the saddle point and the exothermicities of the reactants and products. Furthermore, analytic expressions for the corresponding barrier potential integral are available [41]. Thus, it provides a particularly useful representation for obtaining improved yet simple tunneling estimates.

The combination of Eckart tunneling probabilities with conventional RRKM state counting provides what could be considered as the standard RRKM calculation. Importantly, its implementation requires only energies, rotational constants, and harmonic vibrational frequencies at the saddle point. Meaningfully accurate values for such quantities can often be obtained from quantum chemical simulations employing standard software packages. Coupled cluster calculations employing perturbative triples corrections (either CCSD(T) [42, 43] or QCISD(T)[44]) provide amongst the most accurate results for these quantities for methods that are reasonably feasible to implement [45]. These results are especially accurate when extrapolated to the infinite basis set limit. However, there are cases where such single reference based methods are inappropriate, with a large spin-contamination and/or $\|T1\|$ diagnostic [46] being indicative of likely failures. In such instances, multi-reference

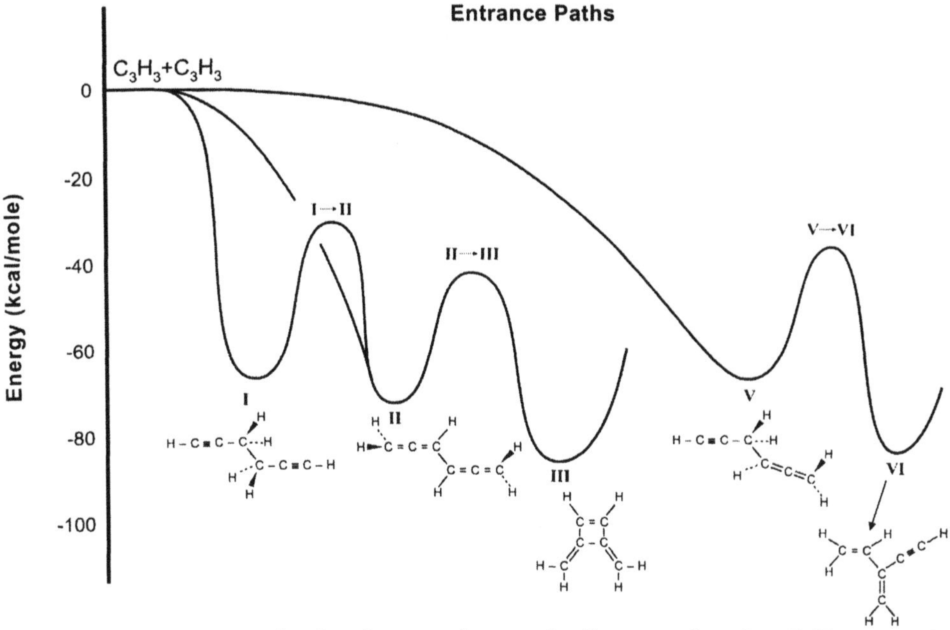

Fig. 2.1. Entrance paths in the reaction path diagram for the C_3H_3 + C_3H_3 reaction.

configuration interaction calculations, which are much more difficult both conceptually and computationally, are generally required. Unfortunately, multi-reference configuration interaction calculations are generally limited in applicability.

For the vibrational frequencies and rotational constants, density functional theory calculations with the relatively modest 6-311G** basis set generally suffice. However, for accurate estimates of the lowest frequency modes it is occasionally necessary to include diffuse functions in the vibrational analyses and so it is perhaps better to employ the 6-311++G** basis. Truhlar and coworkers have recently noted the related importance of diffuse functions in estimating relative conformational energies with density functional theory [47]. For density functional calculations there appears to be little rationale for employing the more expensive correlation-consistent basis sets, since the results are essentially converged with the 6-311++G** basis set and since the limited accuracy of the density functional methods does not warrant extrapolation to the infinite basis set limit.

For wideranging studies of multiple reaction channels conventional RRKM theory with Eckart tunneling provides a readily implemented methodology that generally provides meaningful predictions. Consider for example the self-

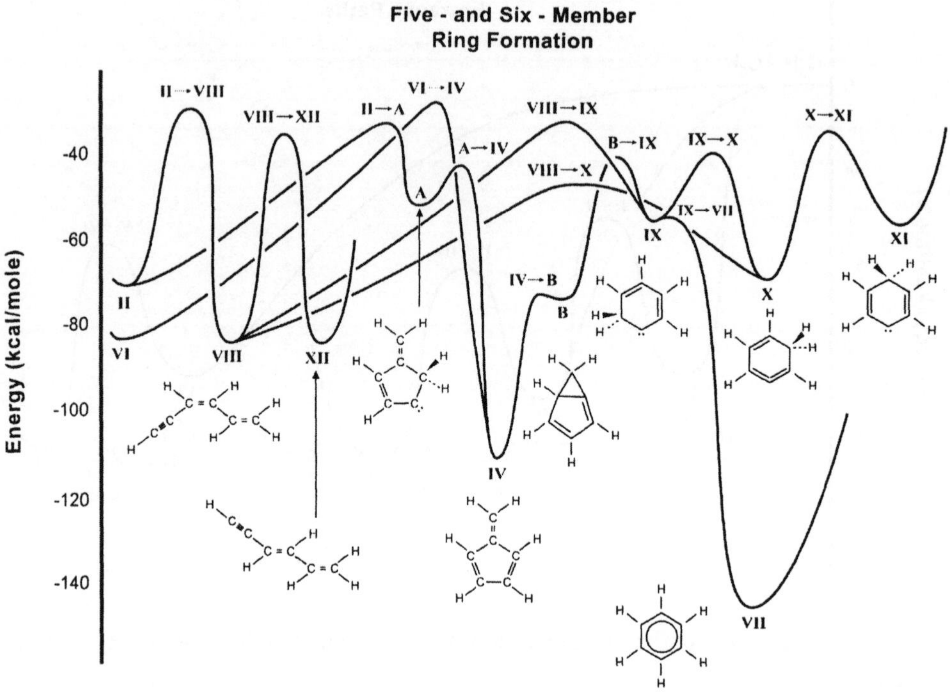

Fig. 2.2. Reaction path diagram for five and six membered ring formation in the $C_3H_3 + C_3H_3$ reaction.

recombination of C_3H_3 radicals. As illustrated in Figs. 2.1-2.4, a current model for this reaction includes 12 wells and 22 transition states [48]. Conventional RRKM calculations with Eckart tunneling provided the basis for a detailed master equation simulation of the kinetics on this surface [48]. The stationary point properties were obtained on the basis of QCISD(T)/6-311G** and MP2(Full)/6-311++G(*3df,2pd*) calculations at B3LYP/6-311++G** optimized geometries. The comparison in Fig. 2.5 of the predicted rate constants (HL1 curve) for the I → III isomerization with the corresponding experimental values [49, 50] illustrates the typical accuracy of such calculations. The solid curve, which is obtained by increasing the I → II barrier height by 1.0 kcal/mol and the II → III barrier height by 0.7 kcal/mol is in remarkably good agreement with the experimental data. In contrast, results based on the earlier BAC-MP4 potential of Miller and Melius [51] are three orders of magnitude too high, while the uncorrected HL1 results are only a factor of two too high. The accuracy of relatively feasible quantum chemical simulations has certainly improved over the last 10 years.

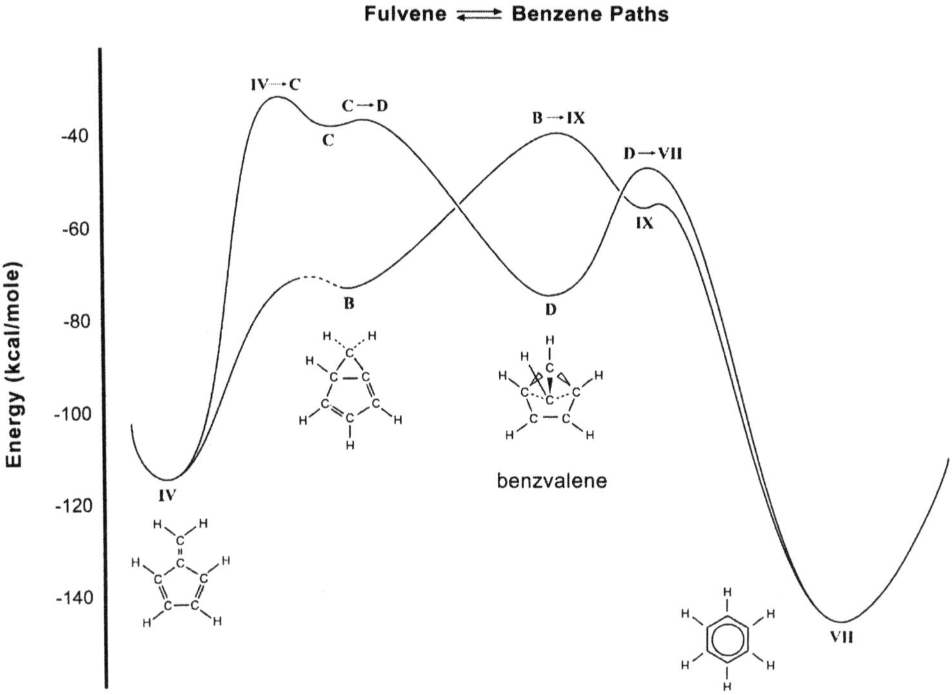

Fig. 2.3. Reaction path diagram for the isomerization of fulvene to benzene.

2.3.2. The reaction path: variational effects and tunneling

For more accurate RRKM predictions, various refinements, such as variational minimization of the transition state flux and multidimensional tunneling corrections, are commonly employed. In variational studies the dividing surfaces have traditionally been specified in terms of a single parameter s, termed the reaction coordinate, given by the distance along a reaction path which connects reactants to products. The "intrinsic" reaction coordinate, which is defined as the distance along the path of steepest descents from the saddle point in a coordinate system for which the kinetic energy expression is diagonal and with constant effective mass, has become the standard reaction coordinate [52-54]. The corresponding reaction path is generally termed the minimum energy path.

A standard reaction path Hamiltonian is derived from a quadratic expansion of the potential in the modes orthogonal to the reaction coordinate [55-59]. For zero total angular momentum this Hamiltonian is expressed as

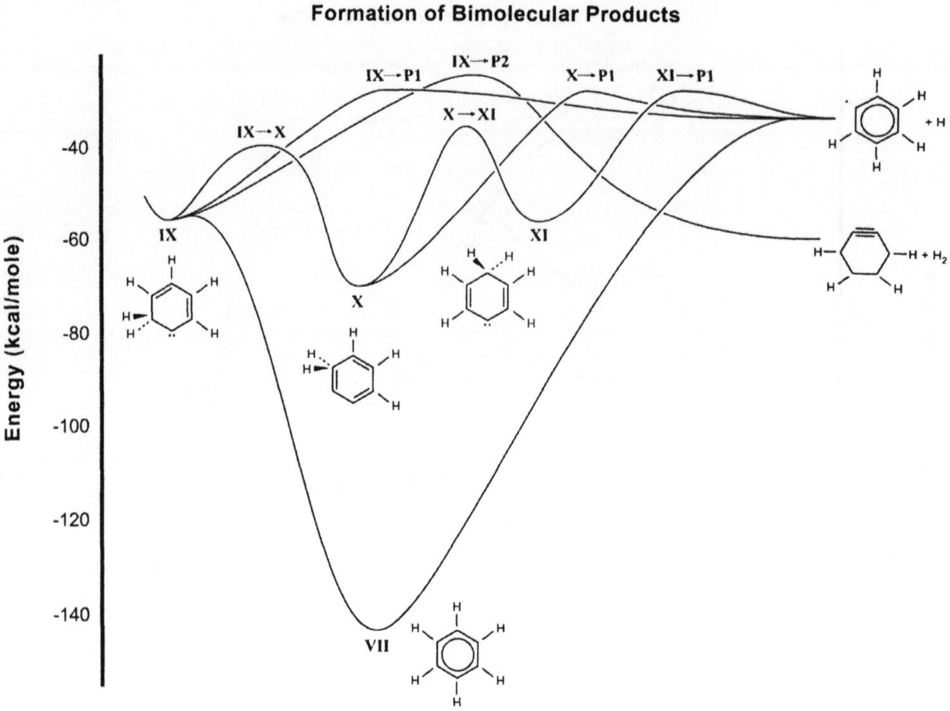

Fig. 2.4. Reaction path diagram for the formation of bimolecular products in the $C_3H_3 + C_3H_3$ reaction.

$$H\left(p_s, s, \{P_i, Q_i\}\right) = \sum_{i=1}^{3n-7} \frac{1}{2}\left(P_i^2 + \omega_i(s) Q_i^2\right) + V_0(s) + \frac{1}{2} \frac{\left[p_s - \sum_{i=1}^{3n-7} Q_k P_l B_{k,l}(s)\right]^2}{\left[1 + \sum_{i=1}^{3n-7} Q_{kl} B_{k,3n-6}(s)\right]^2} . \tag{2.21}$$

The $3n$-7 vibrational frequencies, $\omega_i(s)$, are obtained from normal-mode analyses at points along the reaction path via diagonalization of a projected force constant matrix that removes the translational, rotational, and reaction coordinate motions. The *B* coefficients are defined in terms of the normal mode coefficients, with those in the denominator of the last term determining the reaction path curvature, while those in the numerator are related to the non-adiabatic coupling of different vibrational states. A generalization to non-zero total angular momentum is available [59].

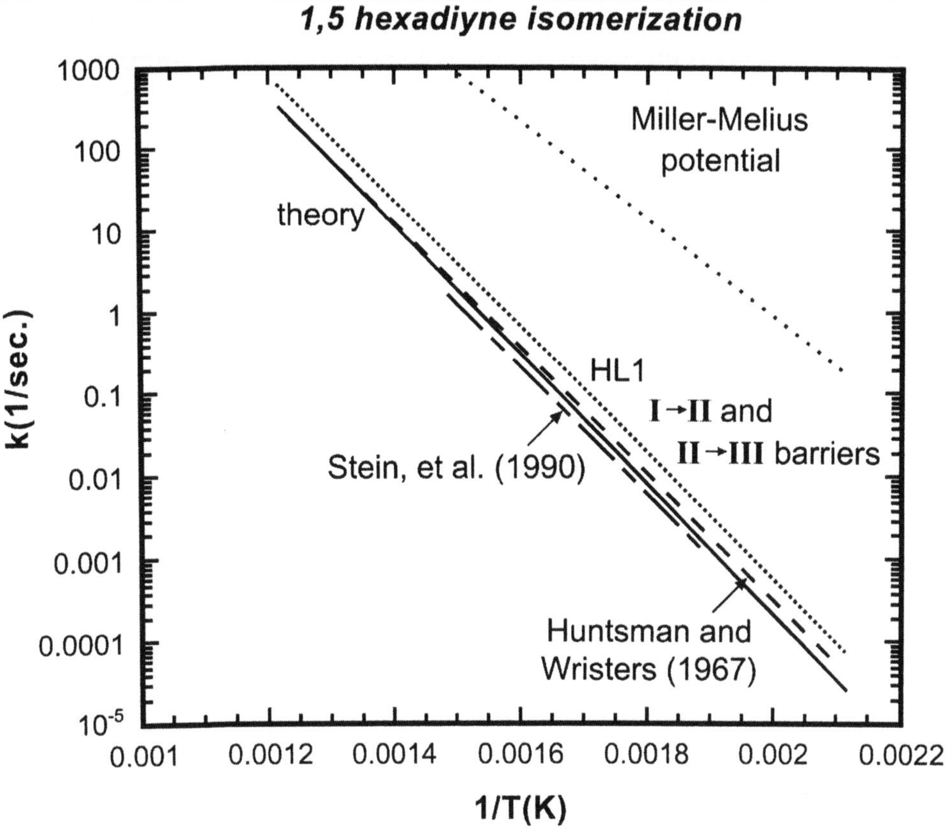

Fig. 2.5. Arrhenius plot of the I → III rate coefficient in C_6H_6.

The simplest, and most commonly implemented, treatment of variational effects involves a neglect of both the non-adiabaticities and the reaction path curvature, which yields simple rigid-rotor harmonic-oscillator energy levels along the reaction path. Procedures for mapping out the reaction path in the intrinsic reaction coordinate, and for determining the projected vibrational frequencies are an integral part of standard quantum chemistry packages. Such computations are generally performed with mass-weighted Cartesian coordinates. Variational effects are typically quite minor (e.g., corrections of 50% or less at temperatures up to 2000 K), especially for reactions with large barriers, and generally increase with temperature. However, variational effects can yield order of magnitude corrections, particularly when the reaction path curvature is large, as is the case with the transfer of a hydrogen atom between two heavy moieties [14]. Thus, due also to the relative ease of incorporating them, the inclusion of variational corrections is quite commonplace.

Furthermore, the potential energy information required for their inclusion is often generated in the process of obtaining improved tunneling estimates.

The determination of accurate quantum tunneling corrections presents the greatest difficulties for *a priori* RRKM calculations due to the non-local nature of tunneling. Truhlar and coworkers have developed a variety of useful schemes for estimating the tunneling correction [15, 60-67]. These schemes incorporate "multidimensional" tunneling paths that deviate dramatically from the reaction path. The appropriate correction formula to apply depends in large part on the extent of the reaction path curvature, with greater curvatures requiring more complex methodologies, considerably greater potential energy information, and larger inherent uncertainties. For small curvatures in the reaction coordinate, Truhlar's small-curvature tunneling (SCT) result [60-63] is a generalization of the early work of Marcus and Coltrin [31] who proposed a vibrationally adiabatic corner-cutting tunneling pathway. For larger curvatures, vibrational adiabaticity breaks down, but it still remains important to search for the most direct, or least action pathways [66, 68-70]. Such considerations lead to the large curvature (LCG) [66] and least-action ground state methods (LAG) [67] of Truhlar and coworkers. Recently, an improved large curvature algorithm was presented to deal with failures due to anharmonic effects in the low energy limit [71].

Amongst these methods, the best approach for a given problem is taken to be that which provides the largest correction. Notably, according to a review [14] "It appears that the practical problem of estimating tunneling contributions, given the potential energy surface, is largely solved, at least for accuracies of a factor of 2 or better at temperatures of 250 K or higher." Furthermore, a recent comparison of multidimensional tunneling calculations with accurate full-dimensional quantum dynamics calculations for $H + CH_4 \rightarrow H_2 + CH_3$ finds a maximum deviation of only 23% [72]. It is beyond the scope of this article to discuss the various schemes in any depth. The reader is referred to the comprehensive summaries provided by Truhlar and coworkers in various places [32, 73].

A recent application of zero-curvature WKB tunneling corrections [74] found significant differences from the corresponding Eckart barrier results. However, the tunneling factor is still most sensitive to the imaginary frequency at the saddle point. Thus, the evaluation of one-dimensional reaction path tunneling probabilities is most useful when the imaginary frequency has itself been accurately determined. Unfortunately, the accurate determination of the imaginary frequency is often quite difficult, requiring large basis set coupled cluster evaluations, for example.

The extreme sensitivity of the predicted rate coefficients to the barrier height, and, to a lesser extent, the barrier frequency, has led to the development of schemes for coupling low level evaluations of the energetics along the reaction path, with higher level evaluations of the properties of the saddle point [75, 76]. The coupling of these "dual-level" methods with the direct determination of the reaction path properties from quantum chemical simulations has proven to be an effective strategy, yielding quantitative results at greatly decreased computational cost. These dual-level direct dynamics methods often employ density functional theory or even fitted semi-empirical evaluations of the reaction path properties.

An interesting illustration of the use of a fitted density functional approach, where the functional is fit to higher level CCSD(T) data, was provided in a recent study of the D + C_2H_2 reaction by Michael et al. [77]. The use of lower-level methods to explore the reaction path is especially valuable for large curvature cases, where additional potential energy surface information is required to obtain accurate tunneling estimates. Related interpolated variational TST methods, which are also proving to be quite effective, focus on interpolating a limited number of high level estimates in place of the low-level high level electronic structure coupling [78-80].

An analysis of the H + C_2H_2 reaction employing a large basis set CCSD(T) analysis of the saddle point and a B3LYP analysis of the harmonic reaction path has recently been completed [81]. The effects of reaction path variational optimizations and Eckart tunneling corrections for the high pressure addition of H to C_2H_2 are illustrated in Fig. 2.6. Variational effects are minimal, reducing the rate constant by 25% or less up to 2000 K. In contrast, the effects of tunneling are quite substantial, increasing the room temperature rate constant by a factor of 2.2. The apparent discrepancy between theory and experiment [82-84] is due to the failure of some of the experiments [82, 84] to reach the high pressure limit, particularly at the higher temperatures [81]. The good overall agreement between theory and experiment obtained in these two studies of the H + C_2H_2 reaction [77, 81] is another testament to the accuracy of current quantum chemical simulations.

One of the practical difficulties encountered in applications involves the difficulty of accurately mapping out the minimum energy path. Accurate determinations necessarily require very small step sizes, and thereby many steps and corresponding quantum chemical simulations to cover a given range. Improved reaction path following algorithms have helped to reduce the number of steps required for a given precision, but the computational requirements can still be excessive [85-88]. A novel procedure, which focuses not on the minimum energy path, but rather on optimal dividing surfaces, largely bypasses these difficulties [89-91]. This procedure considers the optimal orientation of the dividing surface for a given approximate point on the minimum energy path. The minimum on the reoriented dividing surfaces then yields an improved point on the minimum energy path. Furthermore, the normal to the reoriented dividing surface provides a new approximate reaction coordinate. The free energy on the

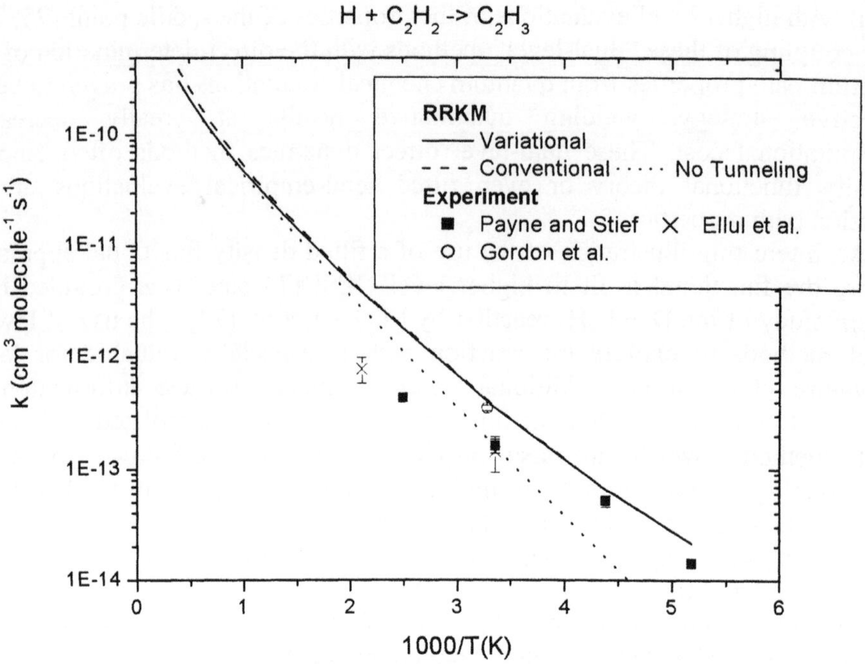

Fig. 2.6. Arrhenius plot of the high pressure addition rate constant for $H + C_2H_2$.

path of reoriented dividing surfaces is found to be much more stable, allowing for significantly larger step sizes.

Other procedures for reducing the computational requirements have focused on generalizations of detailed calculations for one particular reaction to results for related reactions. For example, Truong and coworkers have used reaction path information from the smallest reaction in a given class of reactions to calculate tunneling contributions for larger reactions in the class [92-95]. Similarly, Green and coworkers have generalized group additivity rules to the properties of transition states [96-98]. These methods have been applied initially to hydrogen abstraction reactions.

The *Polyrate* program of Truhlar and coworkers [99] and *TheRate* program of Truong and coworkers [100] provide valuable implementations of many aspects of the above described transition state theory based methodologies for treating the dynamics of reactions with well defined barriers. Particularly valuable in these two programs are their detailed treatments of tunneling corrections and procedures for directly coupling with quantum chemical simulations. The *Multiwell* program of Barker and coworkers [101], the

ChemRate program of Tsang and coworkers [102], and the *VariFlex* program [103] provide more general kinetics programs which include conventional RRKM theory with Eckart or other simple one-dimensional tunneling corrections.

2.3.3. Extensions of RRKM theory

The standard derivation of RRKM theory assumes that the reaction occurs on a single adiabatic electronic state. Generalizations to the treatment of nonadiabatic reactions, involving, for example spin-forbidden transitions, have been presented [104-107]. With these generalizations, the normal mode expansion at the saddle point is replaced by a corresponding expansion about the minimum on the seam of crossing between the two diabatic electronic states. A nonadiabatic transition rate is then obtained by the replacement of the one-dimensional transition probabilities of RRKM theory (i.e., the step functions of Eq. (2.8) or the tunneling probabilities of Eq. (2.10)) with 2 times the 1-dimensional curve crossing probability. The curve crossing transition probabilities are often evaluated within the Landau-Zener model:

$$P_{non-adiabatic}^{Landau-Zener}(E) = 1 - \exp\left(-\frac{2\pi V_{12}^2}{|\mathbf{v}\bullet\Delta\mathbf{F}|\hbar}\right), \qquad (2.22)$$

where V_{12} is the matrix element for the coupling between the two diabatic states, $\mathbf{v}$ is the nuclear velocity perpendicular to the crossing seam, and $\Delta\mathbf{F}$ is the difference in the potential gradients for the two diabatic states, all evaluated at the minimum on the seam of crossing. Alternatively, one can obtain a quantum mechanical solution [107] for the one-dimensional curve crossing probability. Recent applications of nonadiabatic RRKM theory have demonstrated the means for obtaining the necessary parameters directly from quantum chemical simulations [107, 108].

RRKM theory is also not directly applicable to the study of the branching of product channels that occurs due to a bifurcation of channels after a saddle point. A recent study by Knyazev has provided an interesting discussion of the generalization of RRKM theory to such reactions, emphasizing the importance of the potential energy surface ridges that separate the various channels [109]. The bifurcation point on these ridges was used to approximately separate the flux into the two key bimolecular channels in the CH_3 + O reaction. The particular algorithm employed for the determination of the ridges and bifurcation is somewhat ill defined. However, the study does illustrate the importance of potential energy surface ridges and suggests important pathways for further analyses.

The implementation of RRKM theory at the *E* and *J* resolved level is fundamentally more meaningful than corresponding microcanonical and/or canonical implementations since *E* and *J* are strictly conserved quantities in the absence of collisions. The rotational quantum number *K*, which describes the projection of the total angular momentum on a body-fixed *z*-axis, while not strictly a conserved quantity, may be more or less invariant during the time-scale of the reaction. A classical trajectory analysis of the extent of K-mixing induced by Coriolis interactions was provided by Gray and Davis for H_2CO at relatively low levels of vibrational excitation [110]. Hase and coworkers have considered the difference in RRKM predictions for some illustrative reactions when the *K* rotor is treated as adiabatic (conserved) or active (statistically mixed) for both the transition state and for the reactants [111-113]. Notably, with proper state counting, the treatment of the *K* rotor is irrelevant when considering a fixed transition state. Application to the Cl + C_2H_2 reaction found differences of about a factor of 2 in the low pressure thermal rate coefficient between adiabatic and active treatments for the transition state [112]. Recent experimental studies have begun to probe this question for a few barrierless reactions [114-120]. These studies suggest that, while there does appear to be significant *K*-mixing (particularly for the molecular complex), the mixing in the transition state region may not be complete.

2.4. VIBRATIONAL ANHARMONICITIES AND NON-RIGIDITIES

2.4.1. Background

For accurate RRKM predictions it can be important to consider the effect of deviations in the energy levels from their rigid-rotor harmonic-oscillator values. These deviations affect both the reactant and transition state partition functions. Fortunately, while such deviations can be quite substantial, in many instances their effects are greatly ameliorated by certain cancellations of errors. For example, when considering canonical rate coefficients, the contribution of vibrational anharmonicities to the transition state partition function is cancelled to a great extent by the corresponding contribution to the reactant partition function. This is particularly true of the "spectator" modes that are not directly involved in the reaction. For other modes, such as the interfragment bending modes in bond dissociations, the anharmonic effects for the reactant and transition state can be in opposite directions. Such cancellations are especially important for large molecules, where the net effect of anharmonicities on one of the partition functions corresponds to the product of the effects for each of the vibrational modes. This cancellation of errors is likely the reason for the relatively small anharmonic correction (17%) in Isaacson's conventional TST calculations for the $OH + H_2 \rightarrow H_2O + H$ reaction at room temperature [121], when their earlier calculations of the transition state vibrational partition function found anharmonic effects of a factor of 1.7 [122]. Notably, their tunneling-corrected results were much more strongly dependent on anharmonic effects (the correction factor was approximately 8 at room temperature) due to decreases in both the zero-point energy at the saddle point and in the barrier width.

At the microcanonical level, vibrational anharmonicities might be expected to have a greater effect on the reactant density of states than on the transition state number of states, due to the increased available energy for the former quantity. An interesting quantitative study of the effects of vibrational anharmonicities on the microcanonical rate coefficients for the $HCO_2 \rightarrow CO_2 + H$ reaction was provided by Christoffel and Bowman [123]. They found, with variational nonseparable quantum calculations of the vibrational energy levels, that anharmonicities *increased* the reactant state density by about a factor of two. In contrast, they *decreased* the transition state number of states, also by about a factor of two, yielding a net decrease in the RRKM dissociation rate constant of a factor of four. The difference in the direction of the anharmonic corrections at the saddle point as opposed to the reactants is mostly due to the $H...CO_2$ bending modes, which are examples of non-spectator modes in a bond dissociation. Notably, the anharmonic zero-point energies differed from the harmonic normal mode values by a greater amount for the transition state (250 cm^{-1}) than for the molecular complex (130 cm^{-1}), or the separated fragments (50 cm^{-1}). The differences in these zero-point energy changes imply significant differences in the thermal high pressure dissociation and association rate coefficients. Such differences are of comparable magnitude to the uncertainties in quantum chemically estimated barrier heights. Overall, this study suggests the importance of detailed considerations of anharmonic effects for the non-spectator modes in bond dissociations.

The generation of pressure-dependent thermal rate coefficients from such microcanonical rate coefficients involves some averaging over a collision induced distribution. This averaging can once again reduce some of the errors due to the neglect of vibrational anharmonicities. In particular, at fairly high pressures, the distribution function is close to Boltzmann. In the calculation of thermal rate coefficients the errors in the distribution function then cancel with those in the microcanonical rate coefficients, just as in the high pressure limit.

In contrast, in the low pressure limit, the rate coefficient is effectively proportional to the state density at the dissociation threshold and there is no cancellation from the transition state number of states. However, in this instance a different sort of cancellation may occur. In particular, for bond stretching modes the vibrational anharmonicities tend to increase the state density, while for some other modes, such as out-of-plane bends, the anharmonicities tend to decrease the state density. As a result, spectroscopic perturbation theory based calculations employing an ab initio determined quartic force field for CH_2CO yielded an anharmonic vibration correction to the density of states of about 20% at the dissociation threshold [124].

2.4.2. Internal rotors

Internal rotational degrees of freedom are generally not well described as either harmonic vibrators or as free rotors (for which simple analytic expressions are also available [7]). Instead, they transform from the harmonic-oscillator limit at low temperature to the free rotor limit at high temperature. For these degrees of freedom the effect of vibrational anharmonicities tends to be greater and more variable from the reactants to the transition state than for many other degrees of freedom. Thus, considerable effort has been devoted to the development of procedures for properly evaluating the hindered internal rotor contributions, focusing primarily on one-dimensional representations and canonical partition functions. Generalizations to the state density have also been presented and it is now standard practice to implement such corrections for each of the hindered rotors in any RRKM calculation.

One key aspect of the internal rotor analyses involves the determination of the reduced moment of inertia for the rotor. In a series of articles, Pitzer and coworkers derived various expressions for decoupling the internal rotor from the external rotor, including the case of multiple rotors [125-128]. East and Radom have recently provided a useful summary of their key results [129]. Robertson and Wardlaw provide an alternative viewpoint [130]. The subsequent extension of this method by Gang et al. is particularly useful in considering multiple hindered rotors [131-133]. Applications to molecules as large as octane were presented in their latest study [133]. Further extension of their approach may also prove useful in considering the coupling of the internal rotors with the remaining vibrational modes.

Another key aspect of internal rotor analyses involves solving for the quantum partition function for a given hindering potential and reduced moments of inertia. Pitzer and Gwinn's early tabulation for a cosine hindering potential [125] has played a prominent role in many of the numerous approximations to the canonical quantum one-dimensional hindered rotor partition function [134-137]. These results generally also make use of the Pitzer-Gwinn approximation for relating the quantum hindered rotor partition function to the product of the quantum harmonic partition function with the classical ratio of the hindered and harmonic partition functions [125]

$$Q_{hindered}^{quantum} \approx Q_{harmonic}^{quantum} \frac{Q_{hindered}^{classical}}{Q_{harmonic}^{classical}}. \tag{2.23}$$

Recent work has focused on generalizing these results to the more common situation of asymmetric hindering potentials [138, 139].

For RRKM theory it is more important to have a procedure for estimating the density of states for hindered internal rotors. Forst [140], Knyazev [141, 142] and McClurg [143], have separately provided approximate expressions for the density of states via inverse Laplace transforms of the canonical partition functions. We have employed the Pitzer-Gwinn approximation at the microcanonical level [144], with simple configurational integrals providing the classical state densities [145]. Jordan et al. determined analytic classical partition functions and densities of states for a variety of hindering potentials [146]. While each of these expressions is of some utility it is somewhat surprising, with the speed of today's computers, that the standard practice does not involve the simple determination of the quantized, or perhaps the semiclassical, energy levels.

For larger molecules, with multiple internal rotors, it becomes difficult to uniquely distinguish the internal rotors from other low frequency modes. A major advance was provided by Ayala and Schlegel, who provided an automatic identification of the internal rotor modes from the Hessian matrix commonly determined in quantum chemical simulations of the vibrational modes [137]. Their coupling of this procedure with the Kilpatrick and Pitzer protocol [127] for determining the reduced moments of inertia, and an approximate representation of the quantum partition function for cosine hindered potentials, provides a useful automated procedure for estimating the effect of hindered rotors on the canonical partition functions. The treatment of the multiple internal rotors corresponding to the relative bending motions of the two fragments arising in simple bond dissociations is considered in detail in the next section.

For molecules with multiple torsional modes the hindered rotor motions often have significant potential couplings. However, products of one-dimensional treatments are still generally applied due to the difficulty of mapping out the multidimensional potential energy surfaces. Such one-dimensional treatments are especially appealing since reasonable approximations to the hindering potential can be obtained from just the quadratic expansion of the potential about the minimum energy configurations. Alternatively, additional quantum chemical determinations of the maximum in the hindering potential, or complete samplings of the angular dependence of the potential can be used to obtain improved Fourier series representation of the potential. An interesting example of a fully two-dimensional treatment of the hindered internal rotors in the reaction of ethylene with butylbenzene was recently provided by Van Speybroeck and coworkers [147]. For this reaction, the net effect of the potential couplings on the canonical rate coefficients corresponds to a reduction in the rate coefficient by about 30%, due in part to some cancellation of errors in the partition functions for the transition state and the reactants.

The transfer of an H atom from one site to another, as in the HCN → NCH isomerization, can be viewed as a special type of internal rotation. A hindered internal rotor treatment of such motions was found [148, 149] to yield an increase in the reactant state density by a factor of 3 to 4 for both HCN and HCCH at the thresholds for CH bond dissociations. Furthermore, for HCN, where the dissociation energy is well known, the resulting low pressure limit rate coefficients were found to be in much improved agreement with experiment. This study also provided a simple general formula for estimating the effect of such corrections for arbitrary isomerizations (Eq. (2.31) in [149]). Illustrative calculations suggested that such effects may be important even in larger molecules.

2.4.3. Other separable modes

Normal mode analyses are often used to obtain approximately separable degrees of freedom. One-dimensional anharmonic corrections are then easily implemented for these modes via a generalization of the Beyer-Swinehart algorithm [150]. Bond stretching modes are often treated as Morse oscillators, due to the simple analytic expressions for the corresponding quantized energy levels. Furthermore, the anharmonicity parameter can be estimated from the dissociation energy when it is not known from spectroscopy. Bending modes have occasionally been treated with mixed quadratic-quartic potentials [151]. Katzer and Sax presented a procedure for obtaining effective rotational constants for the pseudorotational motions that arise in Jahn-Teller distorted molecules like cyclopentadienyl radical and in puckered cyclic molecules [152]. Knyazev and Tsang have recently generalized their results for internal rotor state densities to obtain an algorithm for deriving approximate quantum anharmonic state densities for arbitrary potential energy forms [153]. The algorithm is based on classical phase space integrals coupled with quantum corrections obtained via the Pitzer-Gwinn approximation and inverse Laplace transforms of the canonical partition functions. Again, it is unclear why the standard practice for one-dimensional anharmonicity corrections does not involve the determination of the semiclassical action on a grid of energies.

2.4.4 Full treatments

While mode decoupling schemes allow for the straightforward implementation of one-dimensional anharmonicity corrections, it is not clear that they can be expected to provide quantitatively accurate corrections. Thus, procedures for considering the effects of mode coupling have been discussed by many groups. At the classical level, accurate fully coupled numbers of states are easily expressed in terms of phase space integrals that can generally be reduced to just configurational integrals [154]. Monte Carlo evaluation provides a standard procedure for evaluating the multidimensional configuration integrals. Densities of states can be obtained via either numerical or analytic differentiation. Again, quantum corrections may be implemented with a microcanonical version of the Pitzer-Gwinn approximation. Alternatively, the usual Pitzer-Gwinn approximation could be applied to the classically evaluated canonical partition functions, followed by inverse Laplace transforms. Parneix and coworkers propose a different approach based on the temperature dependence of the average energy [155].

Such configuration integral based procedures have been difficult to implement due to the need for a potential energy surface for all configurations sampled. Thus, such methods have generally been applied to only triatomic molecules for which analytic potential energy surfaces exist. However, the continuing advance in efficiency of quantum chemical simulations and in generally accessible computational capabilities suggests that direct integration schemes should now be feasible. A preliminary study indicates that such a direct configurational integral based approach is readily applicable to the determination of fully coupled classical state densities for CH_4 [156]. Furthermore, it should be feasible to apply such methods to a wide variety of other 5 and 6 atom molecules, and perhaps significantly larger molecules.

An alternative classical approach involves an adiabatic switching from dynamics on a reference Hamiltonian of known state density to dynamics on the true Hamiltonian for the system of interest [157]. Application of this approach to the study of Al_3 clusters yielded anharmonic correction factors of 2.5 to 2.9 [158]. A related "reference system equilibration" approach has also been proposed [159]. Empirical models for the coupling between bond stretching and angle bending degrees of freedom have been studied [160, 161] and suggest significant effects due to this coupling.

Detailed spectroscopic and kinetic observations for NO_2 have suggested interesting irregularities in the state densities near the dissociation threshold, due to varying contributions from long range separations [162-169]. Long-range contributions to the state density from interfragment modes in the dissociation of CH_2CO were studied with a fully anharmonic direct HF/6-31G* approach [156]. The results obtained for the ratio of the classical anharmonic to classical

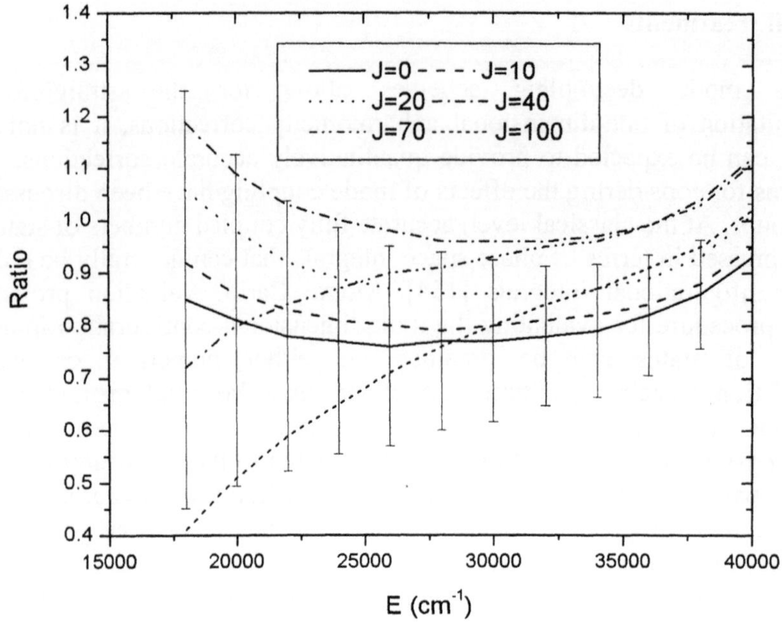

Fig. 2.7. Plot of the ratio of the anharmonic to harmonic state density versus energy relative to the bottom of the well for a range of total angular momentum in CH_2CO considering only the anharmonicities in the intermolecular modes of the CH_2 + CO channel. The classical dissociation threshold is at about 33,000 cm^{-1}. The curve for J=0 includes Monte Carlo integration uncertainty error bars.

rigid rotor harmonic state density are plotted in Fig. 2.7. The finding of a ratio of 0.8 +/-0.2 at the dissociation threshold is in reasonable agreement with the conclusion reached on the basis of spectroscopic perturbation theory. At least for this molecule, the long-range separations make a negligible contribution when the geometries are restricted to be inside of the transition state, i.e., CC separations of 3 Å or less. The approach employed is completely general and has been used in a related MP2 study of the density of states for the interfragment modes in the Cr^+ + C_6H_6 ion-molecule reaction [170]. A recent study of anharmonicities in the Cl^- + CH_3Cl reaction incorrectly suggests that classical approaches cannot be applied to the full dimensional problem due to difficulties with incorporating quantum corrections [171]. In fact, approximate quantum anharmonic results can easily be obtained via the microcanonical

Pitzer-Gwinn approximation, coupled with constraints on the range of bond-lengths to maintain undissociated fragments.

Purely quantum studies of the fully coupled anharmonic (and sometimes nonrigid) rovibrational state densities have also been obtained with a variety of methods. The simplest to implement are spectroscopic perturbation theory based studies [121, 122, 124]. Related semiclassical perturbation treatments have been described by Miller and coworkers [172-174]. Vibrational self-consistent field (SCF) plus configuration interaction (CI) calculations [175, 176] provide another useful alternative, for which interesting illustrative results have been presented by Christoffel and Bowman for the H + CO_2 reaction [123] and by Isaacson for the H_2 + OH reaction [121]. The MULTIMODE code provides a general procedure for implementing such SCF-CI calculations [177]. Numerous studies of the state densities for triatomic molecules have also been presented.

The path integral formalism provides a useful approach for evaluating the fully coupled quantum canonical partition function [178-181]. Notably, a recent displaced points path integral implementation of this approach obtained an integration accuracy of 0.3% with only 5×10^6 samplings for H_2O [181]. This suggests, that 10% accuracy, which would be reasonable for many purposes, could be obtained with only about 5×10^3 samplings. This number of samplings is easily feasible for direct samplings of the potential. Furthermore, the number of required samplings should grow only modestly with increasing molecular size. Interestingly, a related direct path integral simulation of molecular thermochemistry was recently presented by Glaesemann and Fried [182]. The number and density of states can again be obtained via inverse Laplace transforms. Alternatively, Freeman and Doll have described a related procedure for obtaining the density of states from path integral Monte Carlo simulations [183]. More recently, Wadi and Pollak have used short time propagations to derive and apply expressions for the state density in terms of the moments of the Hamiltonian [184]. The latter can again be obtained from appropriate path integral simulations.

2.5. BARRIERLESS REACTIONS

2.5.1. Background

For radical-radical recombination reactions and ion-molecule reactions there is generally no saddlepoint on the potential energy surface for the initial association step. The absence of a saddlepoint in these "barrierless" reactions results in considerable variability in the location of the transition state. In particular, at low energies the optimal transition state generally occurs at large separations where the two reacting fragments are only weakly interacting. With increasing energy, the transition state moves in to separations where the chemical bonding interactions are dominant. Thus, a general treatment of such reactions must employ state counting procedures that are appropriate over a wide range of separations. Furthermore, anharmonic effects for the interfragment modes should be particularly important.

The reaction path Hamiltonian based approaches discussed in Sec. 2.3 have been applied to barrierless reactions (see, e.g., Refs. [185-187]). For these reactions, the reaction path is generally determined by starting at fairly large separations and propagating in to the shorter separations of the transition state [188]. Implementations of the reaction path approach have tended to employ rigid-rotor harmonic-oscillator assumptions with occasional corrections based on separable torsional assumptions. While the rigid-rotor harmonic oscillator assumptions may be valid at short separations, where the interactions are quite strong, at larger separations the interfragment modes transform to free rotors and harmonic vibrator assumptions are clearly inappropriate. Various studies have suggested significant deviations (e.g., factors of two or more) from the harmonic oscillator reaction path limit [189, 190]. The implementation of separable hindered rotor assumptions for some of the modes can alleviate such failures as illustrated in some sample calculations for CH_3 + H, Li^+ + H_2O and Li^+ + $(CH_3)_2O$ reactions [113]. However, when both fragments are polyatomic, it is more difficult to obtain reasonable separable hindered rotor approximations.

2.5.2. Phase space theory, flexible RRKM theory, and the statistical adiabatic channel model

It is difficult to generalize the reaction path approach to obtain a treatment that properly considers the coupled hindered rotor nature of the interfragment modes in barrierless reactions. Thus, much of the recent work in this area has instead approached the problem from the perspective of two interacting fragments. Phase space theory (PST), which assumes only weak spherically symmetric interactions between the fragments, provides a useful reference theory for these developments [191-195]. With this assumption, the rotational quantum numbers for the fragments and for the orbital motions are conserved, and one needs consider only the centrifugal barriers for each quantized rovibrational state of the fragments. With quantum PST the contribution to the number of states from the rotational and orbital modes of the fragment is simply evaluated as a direct sum over the rotational and orbital quantum numbers:

$$N_{\text{PST}}^{\text{transitional}}(E,J)=\sum_{j_1}\sum_{j_2}\sum_{k_1}\sum_{k_2}\sum_{l}\Delta(J,j,l)\Delta(j,j_1,j_2)$$
$$\Theta\left(E-E_{\text{rot}_1}(j_1,k_1)-E_{\text{rot}_2}(j_1,k_1)-E_l^{\dagger}\right), \qquad (2.24)$$

where $E_l^\dagger$ is the maximum in the effective centrifugal potential ($V_{eff}(R) = V(R) + l(l+1)\hbar^2/(2\mu R^2)$, with l the orbital angular momentum, μ the reduced mass, and R the distance between the centers-of-mass of the fragments), E_{rot} is the fragment rotational energy, and Δ denotes the triangle inequality. This contribution is convolved with the density of states for the vibrational modes of the two fragments to yield the transition state number of states

$$N_{\mathrm{PST}}(E,J) = \int_0^E d\varepsilon N_{\mathrm{PST}}^{\mathrm{transitional}}(\varepsilon,J)\,\rho_{vib}^{\mathrm{conserved}}(E-\varepsilon). \tag{2.25}$$

The rotational and orbital modes of the fragments are generally termed the transitional modes, since they transform their character during the reaction process [36]. In contrast, the vibrational modes of the fragments are essentially invariant from the transition state on to separated fragments, and have been termed the conserved modes [36]. In the high pressure limit of association reactions, the contribution from the conserved modes to the transition state partition function cancels with that for the reactants, at least within PST, and so can simply be neglected. Notably, even with more accurate treatments of the transitional modes, the two contributions still cancel under reasonable limiting assumptions [196].

In reality, at interfragment separations of 2 to 4 Å, which is where the transition state typically lies for radical-radical reactions, the interactions between the two fragments are highly nonspherical. As a result, PST, while providing a useful reference theory (particularly for the product rotational state distributions), does not provide especially accurate rate estimates. Flexible RRKM theory was developed to properly treat the asymmetries in the potential via classical phase space integral based evaluations of the transitional mode contributions [36, 197-199]. With this approach the transition state dividing surfaces are defined in terms of a fixed separation R between the centers-of-mass of the two reacting fragments.

The expression for the transitional mode contribution to the canonical transition state partition function in flexible RRKM theory is particularly simple [200]:

$$Q_{transitional}(T,R) = Q_{\mathrm{rot}_1} Q_{\mathrm{rot}_2} \left(\frac{2\pi\mu R^2 k_B T}{\hbar^2} \right)^{3/2} \left\langle \exp(-\beta V(\mathbf{\Omega}_{12},\mathbf{\Omega}_1,\mathbf{\Omega}_2,R)) \right\rangle_{\mathbf{\Omega}_{12},\mathbf{\Omega}_1,\mathbf{\Omega}_2}, \tag{2.26}$$

where Q_{rot_i} denotes the rotational partition function for fragment i, and $\boldsymbol{\Omega}_i$ denotes the Euler angles $(\theta_i, \varphi_i, \chi_i)$ describing the orientation in space of fragment

i, with volume element $d\boldsymbol{\Omega}_i$=($\sin\theta_i d\theta_i d\varphi_i d\chi_i$). Similarly, $\boldsymbol{\Omega}_{12}$ denotes the spherical polar angles describing the absolute orientation of the line-of-centers connecting the centers-of-mass of the two fragments, while $\langle \exp(-\beta V) \rangle_{\Omega_{12},\Omega_1,\Omega_2}$ denotes the average $\int d\boldsymbol{\Omega}_{12} d\boldsymbol{\Omega}_1 d\boldsymbol{\Omega}_2 \exp(-\beta V) / \int d\boldsymbol{\Omega}_{12} d\boldsymbol{\Omega}_1 d\boldsymbol{\Omega}_2$. Given a potential energy surface, V, the evaluation of Q from Eq. (2.26) is straightforward via Monte Carlo sampling over the full range of orientational space [200].

Quantum corrections for the modes orthogonal to the reaction coordinate were evaluated via a Feynmann path integral generalization of Eq. (2.26) [200]. For CH_3 + CH_3, such corrections were found to be of negligible importance, being a few percent or less. Similarly small quantum corrections are expected for other radical-radical or ion-molecule reactions. However, radical molecule reactions, which are occasionally barrierless, or nearly so, would be expected to have more significant quantum corrections, since the weaker attractiveness of the intermolecular potential leads to shorter separation distances at the transition state, and thus larger effective bending frequencies and zero-point energies.

Alternative expressions for $Q_{transitional}(T)$ have been obtained via the consideration of relative orientational coordinates in place of absolute orientational coordinates [201]. In this way the dimension of the configurational integral can be reduced by three. This reduction in dimension offers no computational advantage for Monte Carlo integration schemes, but is of key importance in the quadrature based integration schemes employed in some recent algorithms [103]. The expressions for the microcanonical and E/J resolved number of states are more complex and considerable effort has been devoted to simplifying their evaluation [202-204]. Particular emphasis has been placed on analytic integration over the momenta.

The statistical adiabatic channel model (SACM) [205-209], which was developed somewhat earlier than flexible RRKM theory, provides an alternative approach to handling the transformation from harmonic vibrator to hindered rotor to free rotation for the relative bending motions of the two fragments. This approach considers the maxima in the quantized adiabatic channel energies between reactants and products. The quantized channel energies have generally been estimated via interpolations between the bending levels appropriate at short separation and the hindered and free rotational energy levels appropriate at large separations. These interpolations have typically been implemented in an empirical fashion using a limited number of potential parameters adjusted to obtain agreement with experiment. In contrast, the RRKM based studies have tended to focus on the implementation of the full potential for the transitional modes. Potential based implementations of the SACM are hampered by the difficulty of obtaining quantum energy levels for a sufficient range of energies. Implementations of flexible RRKM theory bypass this difficulty with classical evaluations, which are quantitatively appropriate due to the low frequency nature of the transitional modes.

The proper evaluation of the quantized energy levels within the SACM requires a separable reaction coordinate and thus numerical implementations have implicitly assumed a center-of-mass separation distance for the reaction coordinate, as in flexible RRKM theory. Under certain reasonable limits the underlying adiabatic channel approximation can be shown to be equivalent to the variational RRKM approximations. Thus, the key difference between flexible RRKM theory and the SACM is in the focus on the underlying potential energy surface in flexible RRKM theory as opposed to empirical interpolation schemes in the SACM. Forst's recent implementation of micro-variational RRKM theory [210], which is based on interpolations of product and reactant canonical partition functions, provides what might be considered as an intermediate between these two theories.

2.5.3. Variable reaction coordinate RRKM theory

Flexible RRKM theory and the reaction path Hamiltonian approach take two quite different perspectives in their evaluation of the transition state partition functions. In flexible RRKM theory the reaction coordinate is implicitly assumed to be that which is appropriate at infinite separation and one effectively considers perturbations from the energies of the separated fragments. In contrast, the reaction path Hamiltonian approach considers a perspective that is appropriate for the molecular complex. Furthermore, the reaction path Hamiltonian approach with normal mode vibrations emphasizes the local area of the potential along the minimum energy path, whereas flexible RRKM theory requires a global potential for the transitional modes. One might well imagine that each of these perspectives is more or less appropriate under various conditions.

The variable reaction coordinate (VRC) RRKM approach [34, 37, 211] is intended to encompass both perspectives via the consideration of more general definitions for the transition state dividing surfaces. The same separation of transitional and conserved modes is employed as in PST and flexible RRKM theory and the contribution from the transitional modes is again evaluated via Monte Carlo integration of classical phase space integrals. However, the dividing surfaces are now defined in terms of a fixed separation s between arbitrarily located pivot points in each of the fragments. When these pivot points are located at the centers-of-mass of the fragments then the flexible RRKM approach is recovered. Alternatively, locating them at the atoms involved in the incipient bond should yield an approach that is more closely related to the reaction path Hamiltonian approach. Importantly, the use of classical phase space integral expressions allows for the proper treatment of transitional mode

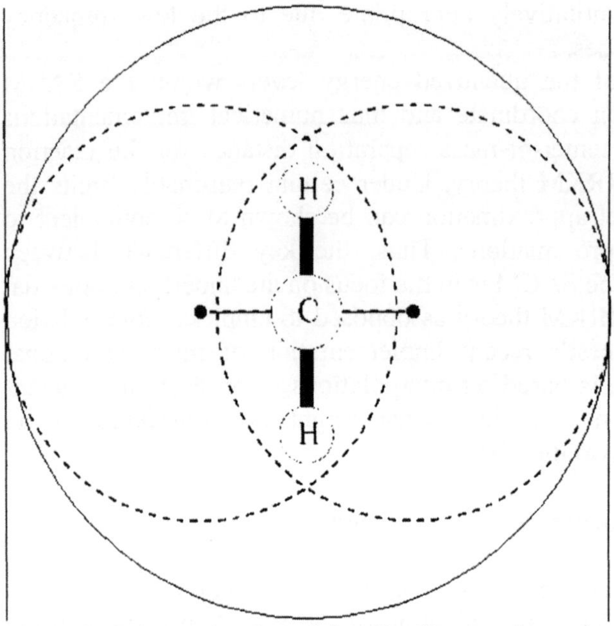

Fig. 2.8. Schematic plot of the dividing surfaces for reaction path (vertical solid line), flexible RRKM theory (solid circle centered about the C atom), and VRC-RRKM theory (dashed circles centered about solid dot pivot points).

vibrational anharmonic and vibration-rotation coupling effects in either limit. Furthermore, various tests have clearly indicated that quantum mechanical effects for the transitional modes are generally of negligible importance [189, 200].

Fig. 2.8 provides a qualitative illustration of the different dividing surfaces considered in the reaction path approach (with Cartesian coordinates), in flexible RRKM theory, and in VRC-RRKM theory for the CH_3 + H reaction [212]. The VRC approach yields a lower rate constant by sampling more repulsive potential values at geometries away from the minimum energy path, e.g., for the incoming H atom approaching one of the CH_3 radical H atoms. Interestingly, detailed optimizations on ab initio potential energy surfaces [213] and model potential based studies of kinetic isotope effects [196], both indicate that the optimal location of the fixed points tends to be within the radical orbitals for the incipient bond. This is illustrated in Fig. 2.9 for the CH_3 + H reaction, where VRC-RRKM predictions [214] for the canonical high pressure addition rate constant are plotted versus temperature for a variety of CH_3 radical pivot point locations. Flexible RRKM theory corresponds to d=0, whereas the optimized VRC-RRKM pivot point is a distance d=0.8 Å away from the C

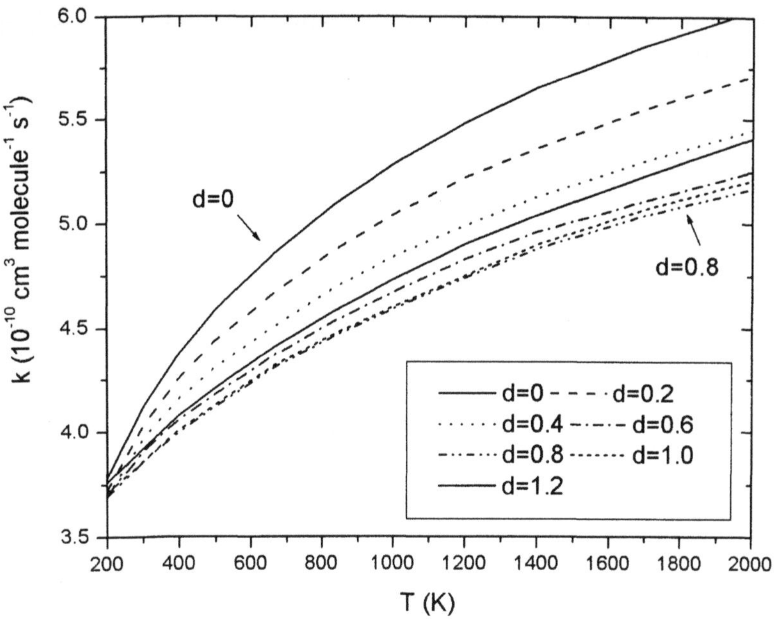

Fig. 2.9. Plot of the canonical VRC-RRKM predictions for the high pressure CH_3 + H addition rate constant for different fixed pivot point locations.

atom. For this reaction, optimizing the reaction coordinate lowers the rate coefficient by only 20% at most. More significant reductions (e.g., factors of 1.5 to 2 are quite typical) are found with larger fragment sizes and/or with two nonlinear fragments.

The initial implementation of the VRC-RRKM approach demonstrated its utility, showing reductions in the computed flux by a factor of two relative to flexible RRKM theory, but required considerable computational resources [34, 37, 211]. The demonstrated utility of the approach led to various efforts to reduce its computational requirements, focusing primarily on analytic integrations over the momenta. The phase space integrals were first reduced to configurational integrals for the canonical and microcanonical cases employing a representation in terms of the absolute orientation of the fragments [34, 215]. Further work by first Smith [216-218], and subsequently by Robertson, Wagner, and Wardlaw [219-221] yielded related expressions in terms of the relative orientations, including results for the *E/J* resolved case. The corresponding reduction in dimension of the configurational integrals has allowed for the implementation of quadrature based schemes in place of Monte Carlo integration [103].

Very recently, a new derivation of the reduction to configurational integrals has been presented for all three cases (canonical, microcanonical, and *E/J* resolved), but with the final configurational integrals once again expressed in absolute orientational coordinates [222]. For the most general case of two nonlinear fragments the canonical partition function is given by

$$Q_{transitional}(T,s)=2\left(\frac{k_BT}{\hbar^2}\right)^4\left(\frac{\mu s^2}{\pi^2}\right)\left\{\prod_{k=1}^{2}\prod_{i=1}^{3}\sqrt{2\pi I_i^{(k)}}\right\}\left\langle\Phi\exp\left(-\beta V\left(\mathbf{\Omega}_{12},\mathbf{\Omega}_1,\mathbf{\Omega}_2\right)\right)\right\rangle_{\mathbf{\Omega}} \tag{2.27}$$

where, $I_i^{(k)}$, is the *i*'th component of the moment of inertia for the *k*'th fragment and the kinematic factor Φ is given by

$$\Phi=\sqrt{1+\mu\sum_{k=1}^{2}\sum_{i=1}^{3}\left(\mathbf{n}^{(12)}\times\mathbf{d}^{(k)}\bullet\mathbf{n}_i^{(k)}\right)^2/I_i^{(k)}}\ . \tag{2.28}$$

In this expression for the kinematic factor, $\mathbf{n}^{(12)}$ is the unit vector pointing from the second pivot point to the first one, $\mathbf{d}^{(k)}$ is the vector connecting the center of mass of the *k*'th fragment to its pivot point, and $\mathbf{n}_i^{(k)}$ is the unit vector directed along the *i*'th principal axis of the k'th fragment. This expression provides a particularly clear indication of the dependence of the kinematic factor on the pivot point displacements.

The related expression for the kinematic factor derived by Smith for the representation in terms of relative orientations [216, 217] is given by

$$\Phi=\left(\frac{\partial R}{\partial s}\right)\left\{\left(\frac{\partial s}{\partial R}\right)^2+\left[\left(\frac{\partial s}{\partial\gamma_{0x}}\right)^2+\left(\frac{\partial s}{\partial\gamma_{0y}}\right)^2\right]\left(\frac{1}{R^2}\right)+\sum_{k=1}^{2}\sum_{i=1}^{3}\left(\frac{\partial s}{\partial\gamma_{i,k}}\right)^2\left(\frac{\mu}{I_i^{(k)}}\right)\right\}^{1/2} .(2.29)$$

where *R* is the distance between the centers-of-mass of the two fragments and the γ's describe the orientations of the line-of-centers and of each of the fragments. Note that this expression involves derivatives of the reaction coordinate relative to the absolute orientation, whereas the configurational integral is based on relative coordinates. Thus, its implementation requires complicated transformations between relative and absolute orientational coordinates. Robertson et al. have derived a related expression given by [219]

$$\Phi = \left\{1 + \mu\left[\nabla R_{\rm cm}{}^{\rm T} \bullet \mathbf{G}_r^0 \bullet \nabla R_{\rm cm}\right]\right\}^{1/2}, \qquad (2.30)$$

where $\nabla R_{\rm cm}$ is the gradient of the center-of-mass separation expressed in terms of the internal and external coordinates, and $\mathbf{G}_{\rm r}{}^0$ is a reduced G-matrix [219-221]. The form of Eq. (2.30) is similar to that in Eq. (2.28), while Smith has demonstrated the equivalence of Eqs. (2.29) and (2.30) [218].

These equivalent expressions clearly demonstrate that the kinematic factor is greater than or equal to unity, and takes its smallest values when the pivot points are located at the fragment centers-of-mass. Thus, the observed reduction in the flux for alternative pivot points arises from reductions in the effective configurational volume, due to higher potential energies away from the minimum energy path. The particular derivation of Φ employed in Ref. [222] allows for the straightforward generalization to alternative dividing surface shapes. Specific results for the case of an elliptical dividing surface and for a planar dividing surface were also presented [222].

Related expressions for the microcanonical and *E/J* resolved transition state partition functions are given by

$$N(E,s) = \left(\frac{\mu s^2}{\Gamma(5)\pi^3\hbar^8}\right)\left\{\prod_{k=1}^{2}\prod_{i=1}^{3}\sqrt{2\pi I_i^{(k)}}\right\} \times\left\langle \Phi \max\left(0, E - V(\mathbf{\Omega}_{12},\mathbf{\Omega}_1,\mathbf{\Omega}_2)\right)^4\right\rangle_{\mathbf{\Omega}}, \qquad (2.31)$$

$$N(E,J,s) = \left(\frac{4J^2\mu s^2}{\Gamma(7/2)\pi^2\hbar^8}\right)\left\{\prod_{k=1}^{2}\prod_{i=1}^{3}\frac{\sqrt{2\pi I_i^{(k)}}}{\sqrt{2\pi I_i(\mathbf{\Omega}_{12},\mathbf{\Omega}_1,\mathbf{\Omega}_2,s)}}\right\} \times\left\langle \Phi\left[\max 0, E - V(\mathbf{\Omega}_{12},\mathbf{\Omega}_1,\mathbf{\Omega}_2,s) - \sum_{i=1}^{3}\frac{J_i^2}{2I_i}(\mathbf{\Omega}_{12},\mathbf{\Omega}_1,\mathbf{\Omega}_2,s)\right]^{5/2}\right\rangle_{\mathbf{\Omega}_J,\mathbf{\Omega}}. \qquad (2.32)$$

The kinematic factor again plays a prominent role in these expressions. Meanwhile, the Boltzmann factor has been replaced with powers of the energy relative to either the potential, or the effective potential, with the latter given by the sum of the potential and the instantaneous rotational kinetic energy. Monte Carlo integration again provides an effective scheme for evaluating these quantities. Importantly, with crude Monte Carlo integration the same potential values may be used to evaluate $N(E,J,s)$ via Eq. (2.32) for the full range of E and J.

2.5.4. Potential energy surfaces in variable reaction coordinate RRKM theory

The implementation of VRC-RRKM theory requires a representation of the transitional mode potential for the full range of transition state locations. For radical-radical reactions this implies that the fragment-fragment interaction energies must be available for arbitrary orientations at separations ranging from about 2 to 4 Å. For ion-molecule reactions the potential is required out to even greater separations, but such long-range values are usually quite well represented by electrostatic expansions. The generation of accurate transitional mode potential energy surfaces for this range of separations presents the greatest difficulty in applying the VRC-RRKM approach. Simple representations involving fits of long-range and bonding forms were used in much of the early work [37, 223, 224], but yield rate constants of uncertain accuracy. Potentials based on fits of simple hindered rotor expansions to the quadratic force constants along the minimum energy path appear to be similarly limited in accuracy to about a factor of two [156, 225].

For more accurate predictions it appears necessary to employ quantum chemical simulations in exploring the full range of the transitional mode potential. For reactions involving an atomic fragment the required potential has only 2 or 3 degrees of freedom. Then it is reasonably feasible to obtain the requisite potential via *ab initio* determinations for a grid of configurations coupled with spline and Fourier series analytic fits. A number of H plus radical reactions have provided valuable test cases for the VRC-RRKM methodology [213, 214, 226, 227]. Analytic potentials for these systems have been generated on the basis of multi-reference configuration interaction calculations, and are thus expected to be highly accurate. These analytic potentials allow for extensive studies of the dependence of the results on the pivot point location, and also allow for comparisons of the results with trajectory simulations [212, 214]. The latter simulations indicate that the VRC-RRKM treatment of the rigid-body capture rate is accurate to within about 10-15%. Interestingly, the variational minimizations of the transition state partition functions indicate that the optimal pivot point tends to be at a distance of about 0.5 to 1.0 Å away from the atom involved in the incipient bond, in the direction of the radical orbital. In effect, the optimal dividing surfaces appear to closely follow the contours of the radical orbital. This knowledge allows for more effective optimizations when the potential energy surface is not as globally available.

A direct statistics approach has proven useful for the cases where both fragments are either linear or nonlinear. With crude Monte Carlo integration the

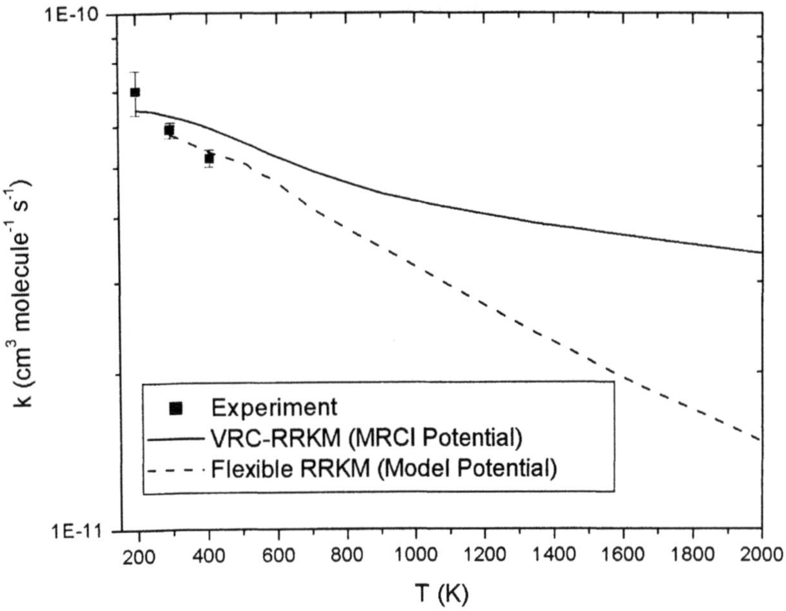

Fig. 2.10. Plot of the CH_3 + CH_3 high pressure recombination rate constants versus T.

configurational integrals in Eq. (2.31) can be determined to within 10% accuracy with about 1000 potential energy surface evaluations. This number of potential energy evaluations is within the limits of feasibility of current quantum chemical simulation methods. Thus, in a number of instances, the generation of analytic potential surfaces was bypassed, and the requisite potential energies were obtained directly from quantum chemical simulations at each of the Monte Carlo sampled configurations [124, 228, 229]. This direct statistics methodology is especially valuable when both fragments are nonlinear and the corresponding analytic potential would be six-dimensional.

The results of a multi-reference configuration interaction (MRCI) based application of the direct statistics method to the CH_3 + CH_3 recombination reaction [228] is illustrated in Fig. 2.10. The predictions are in good agreement with the available low temperature measurements of the high pressure limit [230-232]. At higher temperatures the experiments have difficulty reaching the high pressure limit. The earlier, somewhat empirical, flexible RRKM study of Wagner and Wardlaw [233] can be viewed as a reasonable extrapolation of the higher temperature experimental data to the high pressure limit. Thus, it appears that at higher temperatures the MRCI predictions overestimate the observations by as much as a factor of 2.

Unfortunately, standard single referenced based quantum chemical methods have a qualitative failure in the transition state region for radical-radical reactions. As a result, the QCISD(T) and CCSD(T) methods, which have proven so useful for thermochemical estimates and for reactions with barriers, are of limited use. Instead, accurate predictions have required the use of MRCI methods, whose implementation, unfortunately, tends to be very time-consuming, particularly for larger molecules. Recently, we have been studying the feasibility of employing lower level methods to explore the orientational dependence of the potentials, coupled with one dimensional corrections along the minimum energy path based on multi-reference configuration interaction determinations [234]. Preliminary results suggest that such a dual level scheme is somewhat dependent on the definition of the orientation independence for the 1-dimensional correction. For density functional theory, it appears that the best corrections involve an assumed orientation independence for the optimal dividing surface, or along the potential contours. In contrast, for complete-active space SCF calculations and unrestricted CCSD(T) calculations, the orientation independence is best defined in terms of the distance between the atoms involved in the incipient bond. Two recent quantum chemical developments of coupled cluster methodologies that correct for the standard failure in bond dissociations to two radicals may provide an alternative approach which does not require the implementation of dual level schemes [235-238].

2.5.5. Extensions of variable reaction coordinate RRKM theory

There are some ambiguities and uncertainties in the implementation of VRC-RRKM theory for multichannel reactions that are related to the separation of the flux for the different channels. Multiple addition channel reactions are fairly commonplace, arising in resonantly stabilized radicals where the addition can occur to the radical site in each of the resonance structures, in *p*-orbital radicals where addition can occur from one of two sides, and in ion-molecule reactions where the neutral has multiple electrostatic binding sites. For accurate calculations one should allow the shape of the optimal transition state dividing surface to be different for each channel. Our prior implementations for such reactions have assumed infinite barriers separating the individual channels in order to allow for separate optimizations for each channel [213, 227, 228]. However, the neglect of the flux through the assumed potential barriers can in principle violate the variational principle. In practice that does not seem to be a concern for the reactions to which it was applied, but, in other cases, the separation may not be so clearcut. Robertson et al., in a study of $CFH_2 + H$, give a detailed discussion of such effects [220]. They chose to restrict their consideration to strictly variational dividing surfaces, which likely leads to moderate overestimates in the rate predictions and qualitatively affects the optimal pivot point locations and separations.

The VRC-RRKM approach has recently been generalized to allow for multifaceted dividing surfaces (MDS) composed of individual faces for each separate reaction channel [212]. Consideration of the contributions through each face yields a proper variational result while maintaining flexibility in the form of the dividing surface for each channel. The basis of the MDS approach involves the incorporation of multiple pivot points for each fragment; n_1 for fragment 1 and n_2 for fragment 2. The overall dividing surface is specified in terms of n_1 x n_2 independent transition state separations $s_{i,j}$ between each pair of pivot points by the relation

$$\min\left[\left(\frac{r_{i,j}}{s_{i,j}}\right); i=1,\ldots,n_1; j=1,\ldots,n_2\right] = 1. \tag{2.33}$$

where $r_{i,j}$ is the distance between the i'th pivot point on fragment 1 and the j'th pivot point on fragment 2 for a particular configuration. This approach might be viewed as a generalization of the snowman dividing surface which was independently suggested by Robertson et al. in their study of CFH_2 + H [220]. Sample applications of this MDS-VRC RRKM method have indicated that the optimal multifaceted transition state dividing surfaces are again in qualitative accord with contours of the radical molecular orbital.

The transformation from long-range van der Waals interactions to short range bonding interactions as the reacting fragments approach can lead to the presence of a saddle point below the asymptotic threshold energy. Such a saddlepoint connects a van der Waals minimum with the molecular complex. As a result of this saddlepoint, there are two separate transition state flux minima. One flux minimum, occurring at large separations, beyond the van der Waals minimum, is of key importance at low energies. The other minimum, which occurs at the saddlepoint and shorter separations, is of key importance at higher energies. Notably, the changing form of the interactions can yield two separate transition states even in the absence of a saddlepoint on the potential energy surface, particularly in the *E/J* resolved partition functions.

The process of passing through two separate bottlenecks can reduce the reactive flux below that predicted by the minimum of the two minima. Assuming that the crossing probabilities are given correctly by TST for each encounter of the bottleneck, and that the dynamics is chaotic between the two bottlenecks, implies that the effective transition state flux is given by [19]

$$\frac{1}{N^{\dagger}_{effective}(E,J)} = \frac{1}{N_1^{\dagger}(E,J)} + \frac{1}{N_2^{\dagger}(E,J)} - \frac{1}{N_{\max}(E,J)}, \tag{2.34}$$

where, $N_1^{\dagger}$ and $N_2^{\dagger}$ are the two separate minima in the reactive flux, while $N_{\max}$ is the maximum separating these two minima. Applications of this "unified statistical theory" have helped to explain a number of experimental observations for both neutral and ionic reactions as well as improved the agreement with trajectory simulations. The comparison of theory with photofragment excitation experiments for the singlet dissociation of CH_2CO provides an exceptionally clear demonstration of the effect of the two transition states acting in series [239, 240]. For thermal reactions the averaging over E and J tends to reduce the effects to the point at which they can be ignored (i.e., to about 10% or less).

At higher temperatures the transition state moves in to such short separations that the assumed decoupling of the conserved modes from the reaction coordinate and the transitional modes begins to fail. The optimum internal structure of the fragments begins to deviate from that of the infinitely separated fragments as the separation decreases. Furthermore, the internal vibrational frequencies of the fragments also begin to deviate from those of the free fragments. These variations in the conserved modes are often incorporated in an approximate fashion. An energy relaxation correction, corresponding to the difference in the minimum energy path potential for rigid and relaxed fragment geometries, is often incorporated as a one-dimensional correction. The dynamics in the conserved mode vibrations from the innermost transition state out to separated fragments (or in reverse) are best considered as adiabatic [241, 242]. The changes in their frequencies are thus best incorporated via consideration of the adiabatic potential curves for each vibrational state.

The conserved mode vibrational frequencies tend to increase with decreasing separation due to the increasing interactions between the fragments. Thus, while the energy relaxation correction necessarily increases the predicted rate coefficient, the changes in vibrational frequency tend to offset this increase, but not by as much as the energy relaxation increased it. Recently we have explored the effect of the full orientational dependence for the conserved mode geometry relaxation with direct density functional theory optimizations for the $CH_3 + CH_3$ reaction [243]. The results suggest that the one-dimensional corrections tend to yield an overestimate of the effect of the energy-relaxation, and that this may be a quite general phenomenon. This finding, coupled with the tendency of the geometry relaxation and vibrational frequency variation effects to offset one another further suggests that the overall effect of the conserved mode variations is quite minimal and can often be neglected.

2.6. SUMMARY

The combination of its ease and accuracy has made, and will continue to make, RRKM theory the method of choice for practical predictions of gas phase dissociation and isomerization rate coefficients. The recent and continuing advances in quantum chemical methodologies and computational efficiencies have played an important role in the transformation of RRKM theory from an empirical modeling procedure to a truly predictive theory. The increasing accuracy of the quantum chemical estimates has resulted in numerous illustrations of the quantitative value of *a priori* RRKM calculations. With this new role for RRKM theory it is increasingly important to have implementations that accurately determine the effects of tunneling, of vibrational anharmonicities, of rotational nonrigidities, that employ physically meaningful representations of the transition state dividing surface, and that can be effectively coupled with quantum chemical simulations. To be of practical value these implementations must retain their efficiency, which is centered on the local nature of RRKM theory. In recent years, considerable progress has been made in many of these areas, but there still remain a number of aspects where further developments are needed.

The prediction of low temperature rate coefficients, particularly for reactions with a well defined barrier, continues to be the most difficult task for RRKM theory. Much of this difficulty is related to the exponential dependence of the predictions on the height of the barrier. It is difficult to ascertain the accuracy of current quantum chemical predictions for the barrier heights due to the coupling with uncertainties in the rate constant predictions. Nevertheless, it is clear that the uncertainties in the barrier predictions exceed those for the stable heats of formation, especially when there are significant spin contaminations in the single-reference wavefunctions at the saddle point. Studies of a number of hydrocarbon and oxygenated hydrocarbon reactions suggest that QCISD(T) or CCSD(T) predictions extrapolated to the infinite basis set limit are generally accurate to within about 0.5 to 1.0 kcal/mol, as long as the spin-contamination is minor. In contrast, when the spin-contaminations are large, errors of as much as 4 kcal/mol seem commonplace. Various new methods, such as the spin-flip coupled cluster theory of Krylov and coworkers [235, 236] and the completely renormalized coupled cluster theory of Kowalski and Piecuch [237, 238], hold promise for reducing these errors in the barrier height predictions for spin-contaminated situations, while maintaining the requisite computational feasibility. Alternatively, the range of applicability of multi-reference configuration interaction calculations may increase to the extent where they are widely useful.

The difficulty of accurately modeling the effect of tunneling presents a second bottleneck to making accurate RRKM predictions for the low temperature rate coefficient in reactions with sharp barriers. The essentially nonlocal nature of tunneling makes it difficult to obtain accurate *a priori* tunneling estimates from local information about the saddle point, or even along the reaction path. Thus, continued advances in practical predictions of tunneling probabilities will likely require the introduction of some quantum dynamical considerations. One appealing approach involves the coupling of reduced dimensional quantum dynamics studies of the tunneling motion with vibrationally adiabatic models for the remaining degrees of freedom [244]. The consideration of only a few degrees of freedom with quantum dynamics reduces the computational requirements for the dynamical propagations as well as in the required potential energy surface information.

In some instances, a quantitative understanding of anharmonic effects may be required to acheive *a priori* accuracies of better than a factor of two. Procedures for incorporating one-dimensional corrections, particularly for hindered rotors are well developed and commonly employed. Increased quantum chemical and computational capabilities should now allow for studies of the fully coupled nonrigid anharmonic state densities and/or partition functions via direct Monte Carlo sampling. Such accurate state density studies are a necessary prerequisite to furthering our understanding of the accuracy limits of both quantum chemical estimates and of RRKM theory itself.

Dual level direct dynamics methods have played a key role in practical RRKM predictions for reactions with a barrier. For barrierless reactions, there is a need for improved methods for coupling low-level and high-level simulations, and also for high level simulations that are reasonably feasible for large systems. Again, recent advances in coupled-cluster methodologies may provide appropriate high level simulations of the reaction path [235-238]. One-dimensional corrections based on the atom-atom separations for the incipient bond may provide an appropriate correction to the results from direct CASSCF low level samplings [234]. Alternatively, specific reaction parameterizations (SRP) [78] of density functional methods to fit a high level minimum energy path may yield more consistent predictions.

ACKNOWLEDGEMENTS

Enjoyable and productive discussions on various aspects of the *a priori* implementation of RRKM theory with Yuri Georgievskii, Larry Harding, and Jim Miller are gratefully acknowledged. This research is supported by the Division of Chemical Sciences, Geosciences, and Biosciences, the Office of Basic Energy Sciences of the U.S. Department of Energy. Sandia is a multiprogram laboratory operated by Sandia Corporation, a Lockheed Martin

Company, for the United States Department of Energy's National Nuclear Security Administration under contract DE-AC04-94AL85000.

REFERENCES

[1] R. A. Marcus and O. K. Rice, J. Phys. Colloid. Chem. 55 (1951) 894.
[2] R. A. Marcus, J. Chem. Phys. 20 (1952) 359.
[3] R. A. Marcus, J. Chem. Phys. 43 (1965) 2658.
[4] R. A. Marcus, J. Chem. Phys. 52 (1970) 1018.
[5] W. Forst, Theory of Unimolecular Reactions (Academic, 1971).
[6] P. J. Robinson and K. A. Holbrook, Unimolecular Reactions (Wiley, 1972).
[7] R. G. Gilbert and S. C. Smith, Theory of Unimolecular and Recombination Reactions (Blackwell Scientific, 1990).
[8] K. A. Holbrook, M. J. Pilling and S. H. Robertson, Unimolecular Reactions (John Wiley, 1996).
[9] T. Baer and W. L. Hase, Unimolecular Reactions Dynamics: Theory and Experiment (Oxford, 1996).
[10] J. A. Miller and B. C. Garrett, Int. J. Chem. Kinet. 29 (1997) 275.
[11] J. A. Miller and S. J. Klippenstein, Int. J. Chem. Kinet. 31 (1999) 753.
[12] P. Pechukas, in: Dynamics of Molecular Collisions, Part B, ed: W. H. Miller (Plenum, 1976), p. 269.
[13] P. Pechukas, Annu. Rev. Phys. Chem. 32 (1981) 159.
[14] D. G. Truhlar, W. L. Hase and J. T. Hynes, J. Phys. Chem. 87 (1983) 2664.
[15] D. G. Truhlar, A. D. Isaacson and B. C. Garrett, in: The Theory of Chemical Reaction Dynamics, Vol. IV, ed: M. Baer (Chemical Rubber, 1985), pp. 65.
[16] J. C. Keck, Adv. Chem. Phys. 13 (1967) 85.
[17] Various alternative definitions for χ_R yield equivalent results.
[18] R. A. Marcus, J. Chem. Phys. 41 (1964) 2624.
[19] W. H. Miller, J. Chem. Phys. 65 (1976) 2216.
[20] J. D. Doll, J. Chem. Phys. 73 (1980) 2760.
[21] R. A. Marcus, J. Chem. Phys. 43 (1965) 1598.
[22] W. H. Miller, J. Chem. Phys. 62 (1975) 1899.
[23] F. J. McLafferty and P. Pechukas, Chem. Phys. Lett. 27 (1974) 511.
[24] W. H. Miller, J. Chem. Phys. 61 (1974) 1823.
[25] G. A. Voth, D. Chandler and W. H. Miller, J. Chem. Phys. 91 (1989) 7749.
[26] E. Pollak and D. Proselkov, Chem. Phys. 170 (1993) 265.
[27] N. F. Hansen and H. C. Andersen, J. Chem. Phys. 101 (1994) 6032.
[28] N. F. Hansen and H. C. Andersen, J. Phys. Chem. 100 (1996) 1137.
[29] G. Mills, G. K. Schenter, D. E. Makarov and H. Jonsson, Chem. Phys. 278 (1997) 91.
[30] E. Pollak and J.-L. Liao, J. Chem. Phys. 108 (1998) 2733.
[31] R. A. Marcus and M. E. Coltrin, J. Chem. Phys. 67 (1977) 2609.
[32] D. G. Truhlar and B. C. Garrett, Annu. Rev. Phys. Chem. 35 (1984) 159.
[33] S. K. Gray, S. A. Rice and M. J. Davis, J. Phys. Chem. 90 (1986) 3470.
[34] S. J. Klippenstein, J. Chem. Phys. 94 (1991) 6469.
[35] A. J. Karas and R. G. Gilbert, J. Phys. Chem. 96 (1992) 10893.
[36] D. M. Wardlaw and R. A. Marcus, Adv. Chem. Phys. 70 (1988) 231.
[37] S. J. Klippenstein, J. Chem. Phys. 96 (1992) 367.

[38] D. L. Bunker and W. L. Hase, J. Chem. Phys. 59 (1973) 4621.
[39] T. Beyer and D. F. Swinehart, Commun. ACM 16 (1973) 379.
[40] D. A. McQuarrie, Statistical Mechanics (Harper and Rowe, 1976).
[41] H. S. Johnston, Gas Phase Reaction Rate Theory (Ronald, 1966).
[42] G. D. Purvis and R. J. Bartlett, J. Chem. Phys. 76 (1982) 1910.
[43] K. Raghavachari, G. W. Trucks, J. A. Pople and M. Head-Gordon, Chem. Phys. Lett. 157 (1989) 479.
[44] J. A. Pople, M. Head-Gordon and K. Raghavachari, J. Chem. Phys. 87 (1987) 5968.
[45] It appears to be generally believed that the CCSD(T) method is superior to the QCISD(T) method. However our own experience is that at least for single bond breaking the QCISD(T) method provides a satisfactory solution out to much greater bond separations than does the CCSD(T) method. Our unpublished sample comparisons of heats of formation for a wide range of pure hydrocarbon and oxygenated hydrocarbons suggest little difference otherwise. For this reason, we prefer the QCISD(T) method.
[46] T. J. Lee and P. R. Taylor, Int. J. Qunatum Chem. Symp. 23 (1989) 199.
[47] B. J. Lynch, Y. Zhao and D. G. Truhlar, J. Phys. Chem. A 107 (2003) 1384.
[48] J. A. Miller and S. J. Klippenstein, J. Phys. Chem. A 107 (2003) in press.
[49] S. E. Stein, J. A. Walker, M. M. Suryan and A. Fahr, Proc. Combust. Inst. 23 (1990) 85.
[50] W. D. Hunstman and H. J. Wristers, J. Am. Chem. Soc. 89 (1967) 342.
[51] J. A. Miller and C. F. Melius, Combust. Flame 91 (1992) 21.
[52] A. Kuppermann and D. G. Truhlar, J. Am. Chem. Soc. 93 (1971) 1840 .
[53] H. F. Schaefer, Chem. Br. 11 (1975) 227.
[54] K. Fukui, S. Kato and H. Fujimoto, J. Am. Chem. Soc. 97 (1975) 1.
[55] R. A. Marcus, J. Chem. Phys. 45 (1966) 4493.
[56] R. A. Marcus, J. Chem. Phys. 45 (1966) 4500.
[57] R. A. Marcus, J. Chem. Phys. 49 (1968) 2610.
[58] R. A. Marcus, J. Chem. Phys. 49 (1968) 2617.
[59] W. H. Miller, N. C. Handy and J. E. Adams, J. Chem. Phys. 72 (1980) 99.
[60] B. C. Garrett and D. G. Truhlar, J. Phys. Chem. 83 (1979) 200.
[61] B. C. Garrett and D. G. Truhlar, J. Phys. Chem. 83 (1979) 1079.
[62] B. C. Garrett and D. G. Truhlar, Proc. Natl. Acad. Sci. 76 (1979) 4755.
[63] B. C. Garrett and D. G. Truhlar, J. Chem. Phys. 72 (1980) 3460.
[64] R. T. Skodje, D. G. Truhlar and B. C. Garrett, J. Phys. Chem. 85 (1981) 3019.
[65] R. T. Skodje, D. G. Truhlar and B. C. Garrett, J. Chem. Phys. 77 (1982) 5955.
[66] B. C. Garrett, D. G. Truhlar, A. F. Wagner and T. H. Dunning, J. Chem. Phys. 78 (1983) 4400.
[67] B. C. Garrett and D. G. Truhlar, J. Chem. Phys. 79 (1983) 4931.
[68] V. K. Babamov and R. A. Marcus, J. Chem. Phys. 74 (1981) 1790.
[69] V. K. Babamov, V. Lopez and R. A. Marcus, Chem. Phys. Lett. 101 (1983) 507.
[70] V. K. Babamov, V. Lopez and R. A. Marcus, J. Chem. Phys. 78 (1983) 5621.
[71] A. Fernandez-Ramos and D. G. Truhlar, J. Chem. Phys. 114 (2001) 1491.
[72] J. Z. Pu and D. G. Truhlar, J. Chem. Phys. 117 (2002) 1479.
[73] D. G. Truhlar, B. C. Garrett and S. J. Klippenstein., J. Phys. Chem. 100 (1996) 12771.
[74] J. P. Senosiain, C. B. Musgrave and D. M. Golden, J. Phys. Chem. A 105 (2001) 1669.
[75] W. P. Hu, Y. P. Liu and D. G. Truhlar, J. Chem. Soc. Faraday Trans. 90 (1994) 1715.
[76] Y.-Y. Chuang, J. Phys. Chem. A 101 (1997) 3808.

[77] J. V. Michael, M.-C. Su, J. W. Sutherland, L. B. Harding and A. F. Wagner, J. Phys. Chem. A submitted (2003).
[78] A. Gonzalez-Lafont, T. N. Truong and D. G. Truhlar, J. Chem. Phys. 95 (1991) 8875.
[79] J. C. Corchado, E. L. Coitino, Y.-Y. Chuang, P. L. Fast and D. G. Truhlar, J. Phys. Chem. A 102 (1998) 2424.
[80] A. Fernandez-Ramos, D. G. Truhlar, J. C. Corchado and J. Espinosa-Garcia, J. Phys. Chem. A 106 (2002) 4957.
[81] J. A. Miller and S. J. Klippenstein, J. Phys. Chem. A xxx (2003) to be submitted.
[82] W. A. Payne and L. J. Stief, J. Chem. Phys. 64 (1976) 1150.
[83] E. B. Gordon, B. I. Ivanov, A. P. Perminov and V. E. Balalaev, Chem. Phys. 35 (1978) 79.
[84] R. Ellul, P. Potzinger, R. Reimann and P. Camilleri, Ber. Bunsen. Phys. Chem. 85 (1981) 407.
[85] M. Page and J. W. McIver, J. Chem. Phys. 88 (1987) 922.
[86] C. Gonzalez and H. B. Schlegel, J. Chem. Phys. 90 (1989) 2154.
[87] C. Gonzalez and H. B. Schlegel, J. Phys. Chem. 94 (1990) 5523.
[88] J.-Q. Sun and K. Ruedenberg, J. Chem. Phys. 101 (1994) 2157.
[89] J. Villa and D. G. Truhlar, Theor. Chem. Acc. 97 (1997) 317.
[90] P. L. Fast and D. G. Truhlar, J. Chem. Phys. 109 (1998) 3721.
[91] A. Gonzalez-Lafont, J. Villa, J. M. Lluch, J. Bertran, R. Steckler and D. G. Truhlar, J. Phys. Chem. A 102 (1998) 3420.
[92] T. N. Truong, W. T. Duncan and M. Tirtowidjojo, Phys. Chem. Chem. Phys. 1 (1999) 1061.
[93] T. N. Truong, D. K. Maity and T. T. T. Truong, J. Chem. Phys. 112 (2000) 24.
[94] T. N. Truong, J. Chem. Phys. 113 (2000) 4957.
[95] S. W. Zhang and T. N. Truong, J. Phys. Chem. A 107 (2003) 1138.
[96] R. Sumathi, H.-H. Carstensen and W. H. Green, J. Phys. Chem. A 105 (2001) 8969.
[97] R. Sumathi, H.-H. Carstensen and W. H. Green, J. Phys. Chem. A 105 (2001) 6910.
[98] R. Sumathi, H. H. Carstensen and W. H. Green, J. Phys. Chem. A 106 (2002) 5474.
[99] POLYRATE 9.1: *A New Version of a Computer Program for the Calculation of Chemical Reaction Rates for Polyatomics*, J. C. Corchado, Y.-Y. Chuang, P. L. Fast, J. Villà, W.-P. Hu, Y.-P. Liu, G. C. Lynch, K. A. Nguyen, C. F. Jackels, V. S. Melissas, B. J. Lynch, I. Rossi, E. L. Coitiño, A. Fernandez-Ramos, J. Pu, T. V. Albu, R. Steckler, B. C. Garrett, A. D. Isaacson, and D. G. Truhlar; http://comp.chem.umn.edu/polyrate/.
[100] W. T. Duncan, R. L. Bell and T. N. Truong, J. Comp. Chem. 19 (1998) 1039.
[101] MultiWell-1.3.1, J. R. Barker, (2003); http://aoss.engin.umich.edu/multiwell/; J. R. Barker, Int. J. Chem. Kinetics, 33, 232-45 (2001)
[102] ChemRate, *A Calculational Data Base for Unimolecular Reaction*, V. Mokrushin, W. Tsang, NIST; http://www.nist.gov/kinetics/chemrate/chemrate.html
[103] VariFlex, S. J. Klippenstein, A. F. Wagner, S. H. Robertson, R. C. Dunbar, and D. M. Wardlaw (1999); http://chemistry.anl.gov/chem-dyn/VariFlex/
[104] G. E. Zahr, R. K. Preston and W. H. Miller, J. Chem. Phys. 62 (1975) 1127.
[105] J. C. Lorquet and B. Leyh-Nihant, J. Phys. Chem. 92 (1988) 4778.
[106] A. J. Marks and D. L. Thompson, J. Chem. Phys. 96 (1992) 1911.
[107] Q. Cui, K. Morokuma, J. M. Bowman and S. J. Klippenstein, J. Chem. Phys. 110 (1999) 9469.
[108] A. J. Marks, J. Chem. Phys. 114 (2001) 1700.
[109] V. D. Knyazev, J. Phys. Chem. A 106 (2002) 8741.

[110] S. K. Gray and M. J. Davis, J. Chem. Phys. 90 (1989) 5420.
[111] L. Zhu and W. L. Hase, Chem. Phys. Lett. 175 (1990) 117.
[112] L. Zhu, W. Chen, W. L. Hase and E. W. Kaiser, J. Phys. Chem. 97 (1993) 311.
[113] E. E. Aubanel, D. M. Wardlaw, L. Zhu and W. L. Hase, Int. Rev. Phys. Chem. 10 (1991) 249.
[114] S. W. North and G. E. Hall, Ber. Bunsen. Phys. Chem. 101 (1997) 459.
[115] W. S. McGivern and S. W. North., J. Am. Chem. Soc. 124 (2002) 14472.
[116] W. S. McGivern and S. W. North, J. Chem. Phys. 116 (2002) 7027.
[117] A. Mellinger, M. A. Hall, E. A. Wade and C. B. Moore, Ber. Bunsen. Phys. Chem. 101 (1997) 465.
[118] A. Mellinger, M. V. Ashikhmin and C. B. Moore, J. Chem. Phys. 108 (1998) 8944.
[119] S. Y. Grebenshchikov, H. Flothmann, R. Schinke, I. Bezel, C. Wittig and S. Kato, Chem. Phys. Lett. 285 (1998) 410.
[120] I. Bezel, P. Ionov and C. Wittig, J. Chem. Phys. 111 (1999) 9267.
[121] A. D. Isaacson, J. Chem. Phys. 107 (1997) 3832.
[122] A. D. Isaacson and S.-C. Hung, J. Chem. Phys. 101 (1994) 3928.
[123] K. M. Christoffel and J. M. Bowman, J. Phys. Chem. A 103 (1999) 3020.
[124] S. J. Klippenstein, A. L. L. East and W. D. Allen, J. Chem. Phys. 105 (1996) 118.
[125] K. S. Pitzer and W. D. Gwinn, J. Chem. Phys. 10 (1942) 428.
[126] K. S. Pitzer, J. Chem. Phys. 14 (1946) 239.
[127] J. E. Kilpatrick and K. S. Pitzer, J. Chem. Phys. 17 (1949) 1064.
[128] D. R. Herschbach, H. S. Johnston, K. S. Pitzer and R. E. Powell, J. Chem. Phys. 25 (1956) 736.
[129] A. L. L. East and L. Radom, J. Chem. Phys. 106 (1997) 6655.
[130] S. H. Robertson and D. M. Wardlaw, Chem. Phys. Lett. 199 (1992) 391.
[131] J. Gang, M. J. Pilling and S. H. Robertson, J. Chem. Soc. Faraday Trans. 92 (1996) 3509.
[132] J. Gang, M. J. Pilling and S. H. Robertson, J. Chem. Soc. Faraday Trans. 93 (1997) 1481.
[133] J. Gang, M. J. Pilling and S. H. Robertson, Chem. Phys. Lett. 231 (1998) 183.
[134] D. G. Truhlar, J. Comp. Chem. 12 (1991) 266.
[135] W. Witschel and C. Hartwigsen, Chem. Phys. Lett. 273 (1997) 304.
[136] R. B. McClurg, R. C. Flagan and W. A. Goddard, J. Chem. Phys. 106 (1997) 6675.
[137] P. Y. Ayala and H. B. Schlegel, J. Chem. Phys. 108 (1998) 2314.
[138] Y.-Y. Chuang and D. G. Truhlar, J. Chem. Phys. 112 (2000) 1221.
[139] G. Katzer and A. F. Sax, Chem. Phys. Lett. 368 (2003) 473.
[140] W. Forst, J. Comp. Chem. 17 (1996) 954.
[141] V. D. Knyazev, I. A. Dubinsky, I. R. Slagle and D. Gutman, J. Phys. Chem. 98 (1994) 5279.
[142] V. D. Knyazev, J. Phys. Chem. A 102 (1998) 3916.
[143] R. B. McClurg, J. Chem. Phys. 208 (1998) 1748.
[144] J. D. Doll, Chem. Phys. Lett. 72 (1980) 139.
[145] J. A. Miller and S. J. Klippenstein, J. Phys. Chem. A 104 (2000) 7525.
[146] M. J. T. Jordan, S. C. Smith and R. G. Gilbert, J. Phys. Chem. 95 (1991) 8685.
[147] V. V. Speybroeck, D. V. Neck and M. Waroquier, J. Phys. Chem. A 106 (2002) 8945.
[148] A. F. Wagner, J. H. Kiefer and S. S. Kumaran, Proc. Comb. Inst. 24 (1992) 613.
[149] J. H. Kiefer, P. S. Mudipalli, A. F. Wagner and L. Harding, J. Chem. Phys. 105 (1996) 8075.
[150] S. E. Stein and B. S. Rabinovitch, J. Chem. Phys. 58 (1973) 2438.

[151] B. C. Garrett and D. G. Truhlar, J. Phys. Chem. 83 (1979) 1915.
[152] G. Katzer and A. F. Sax, J. Chem. Phys. 117 (2002) 8219.
[153] V. D. Knyazev and W. Tsang, J. Phys. Chem. A 102 (1998) 9167.
[154] S. C. Farantos, J. N. Murrell and J. C. Hajduk, Chem. Phys. 68 (1982) 109.
[155] P. Parneix, N.-T. Van-Oanh and P. Brechignac, Chem. Phys. Lett. 357 (2002) 78.
[156] S. J. Klippenstein, unpublished results.
[157] W. P. Reinhardt, J. Mol. Struct. 223 (1990) 157.
[158] G. H. Peslherbe and W. L. Hase, J. Chem. Phys. 101 (1994) 8535.
[159] L. E. B. Borjesson, S. Nordholm and L. L. Andersson, Chem. Phys. Lett. 186 (1991) 65.
[160] L. B. Bhuiyan and W. L. Hase, J. Chem. Phys. 78 (1983) 5052.
[161] J. Troe, Chem. Phys. 190 (1995) 381.
[162] R. Jost, J. Nygard, A. Pasinski and A. Delon, J. Chem. Phys. 105 (1996) 1287.
[163] A. Delon, S. Heilliette and R. Jost, Chem. Phys. 238 (1998) 465.
[164] I. Bezel, D. Stolyarov and C. Wittig, J. Phys. Chem. A 103 (1999) 10268.
[165] A. Delon, F. Reiche, B. Abel, S. Y. Grebenshchikov and R. Schinke, J. Phys. Chem. A 104 (2000) 10374.
[166] S. Heilliette, A. Delon, P. Dupre and R. Jost, Phys. Chem. Chem. Phys. 3 (2001) 2268.
[167] S. Heilliette, A. Delon, R. Jost, S. Y. Grebenshchikov, R. Schinke, B. Abel and J. C. Rayez, Zeit. Phys. Chem. 215 (2001) 1069.
[168] D. Stolyarov, E. Polyakova, I. Bezel and C. Wittig, Chem. Phys. Lett. 358 (2002) 71.
[169] S. Heilliette, A. Delon, D. Reignier, T. Stocklin and J. C. Rayez, Phys. Chem. Chem. Phys. 5 (2003) 2039.
[170] C. N. Yang and S. J. Klippenstein, J. Phys. Chem. A 103 (1999) 1094.
[171] A. Delon, S. Heilliette and S. Schmatz, Phys. Chem. Chem. Phys. 4 (2002) 6006.
[172] W. H. Miller, R. Hernandez, N. C. Handy, D. Jayatilaka and A. Willetts, Chem. Phys. Lett. 172 (1990) 62.
[173] M. J. Cohen, N. C. Handy, R. Hernandez and W. H. Miller, Chem. Phys. Lett. 192 (1992) 407.
[174] M. J. Cohen, A. Willetts and N. C. Handy, J. Chem. Phys. 99 (1993) 5885.
[175] K. M. Christoffel and J. M. Bowman, Chem. Phys. Lett. 85 (1982) 220.
[176] H. Romanowski and J. M. Bowman, Chem. Phys. Lett. 110 (1984) 235.
[177] S. Carter, J. M. Bowman and N. C. Handy, Theor. Chem. Acc. 100 (1998) 191.
[178] R. Q. Topper and D. G. Truhlar, J. Chem. Phys. 97 (1992) 3647.
[179] J. Srinivasan, Y. L. Volobuev, S. L. Mielke and D. G. Truhlar, Comp. Phys. Comm. 128 (2000) 446.
[180] S. L. Mielke and D. G. Truhlar, J. Chem. Phys. 114 (2001) 621.
[181] S. L. Mielke and D. G. Truhlar, J. Chem. Phys. 115 (2001) 652.
[182] K. R. Glaesemann and L. E. Fried, J. Chem. Phys. 118 (2003) 1596.
[183] D. L. Freeman and J. D. Doll, J. Chem. Phys. 101 (1994) 848.
[184] H. Wadi and E. Pollak, J. Chem. Phys. 110 (1999) 8246.
[185] W. L. Hase, J. Chem. Phys. 64 (1976) 2442.
[186] S. L. Mondro, S. V. Linde and W. L. Hase, J. Chem. Phys. 84 (1986) 3783.
[187] W. L. Hase, S. L. Mondro, R. J. Duchovic and D. M. Hirst, J. Am. Chem. Soc. 109 (1987) 2916.
[188] W. L. Hase and D. M. Wardlaw, in: Bimolecular Collisions, eds J. E. Baggott and M. N. R. Ashfold (Burlington House, 1989), p. 171.
[189] J. Yu and S. J. Klippenstein, J. Phys. Chem. 95 (1991) 9882.

[190] A. F. Wagner, L. B. Harding, S. H. Robertson and D. M. Wardlaw, Ber. Bunsen. Phys. Chem. 101 (1997) 391.

[191] P. Pechukas and J. C. Light, J. Chem. Phys. 42 (1965) 3281.

[192] P. Pechukas, R. Rankin and J. C. Light, J. Chem. Phys. 44 (1966) 794.

[193] E. E. Nikitin., Teor. Eksp. Khim. Acad. Nauk. Ukr. SSR 1 (1965) 428.

[194] W. J. Chesnavich and M. T. Bowers, J. Am. Chem. Soc. 99 (1977) 1705.

[195] W. J. Chesnavich and M. T. Bowers, J. Chem. Phys. 66 (1977) 2306.

[196] C. A. Taatjes and S. J. Klippenstein, J. Phys. Chem. A 105 (2001) 8567.

[197] D. M. Wardlaw and R. A. Marcus, Chem. Phys. Lett. 110 (1984) 230.

[198] D. M. Wardlaw and R. A. Marcus, J. Chem. Phys. 83 (1985) 3462.

[199] D. M. Wardlaw and R. A. Marcus, J. Phys. Chem. 90 (1986) 5383.

[200] S. J. Klippenstein and R. A. Marcus, J. Chem. Phys. 87 (1987) 3410.

[201] S. H. Robertson, A. F. Wagner and D. M. Wardlaw, J. Chem. Phys. 103 (1995) 2917.

[202] S. C. Smith, J. Chem. Phys. 95 (1991) 3404.

[203] S. C. Smith, J. Chem. Phys. 97 (1992) 2406.

[204] S. C. Smith, J. Phys. Chem. 97 (1993) 7034.

[205] M. Quack and J. Troe, Ber. Bunsen. Phys. Chem. 78 (1974) 240.

[206] M. Quack and J. Troe, Ber. Bunsen. Phys. Chem. 79 (1975) 469.

[207] J. Troe, J. Chem. Phys. 75 (1981) 226.

[208] J. Troe, J. Chem. Phys. 79 (1983) 6017.

[209] J. Troe, Chem. Phys. Lett. 122 (1985) 425.

[210] W. Forst, J. Phys. Chem. 95 (1991) 3612.

[211] S. J. Klippenstein, Chem. Phys. Lett. 170 (1990) 71.

[212] Y. Georgievskii and S. J. Klippenstein, J. Phys. Chem. A 107 (2003) in press.

[213] S. J. Klippenstein and L. B. Harding, Phys. Chem. Chem. Phys. 1 (1999) 989.

[214] S. J. Klippenstein, Y. Georgievskii and L. B. Harding, Proc. Comb. Inst. 29 (2002) 1229.

[215] S. J. Klippenstein, Chem. Phys. Letts. 214 (1993) 418.

[216] S. C. Smith, J. Phys. Chem. 98 (1994) 6496.

[217] S. C. Smith, J. Chem. Phys. 111 (1999) 1830.

[218] S. C. Smith, J. Phys. Chem. A 104 (2000) 10489.

[219] S. H. Robertson, A. F. Wagner and D. M. Wardlaw, J. Chem. Phys. 113 (2000) 2648.

[220] S. Robertson, A. F. Wagner and D. M. Wardlaw, J. Phys. Chem. A 106 (2002) 2598.

[221] S. H. Robertson, D. M. Wardlaw and A. F. Wagner, J. Chem. Phys. 117 (2002) 593.

[222] Y. Georgievskii and S. J. Klippenstein, J. Chem. Phys. 118 (2003) 5442.

[223] S. J. Klippenstein, S. Kristyan, D. L. Yang, T. Yu and M. C. Lin, J. Phys. Chem. A 102 (1998) 6973.

[224] M. Pesa, M. J. Pilling, S. H. Robertson and D. M. Wardlaw, J. Phys. Chem. A 102 (1998) 8526.

[225] S. J. Klippenstein and Y.-W. Kim, J. Chem. Phys. 99 (1993) 5790.

[226] L. B. Harding and S. J. Klippenstein, Proc. Comb. Inst. 27 (1998) 151.

[227] S. J. Klippenstein and L. B. Harding, Proc. Comb. Inst. 28 (2000) 1503.

[228] S. J. Klippenstein and L. B. Harding, J. Phys. Chem. A 103 (1999) 9388.

[229] D.-C. Fang, L. B. Harding, S. J. Klippenstein and J. A. Miller, Faraday Disc. 119 (2001) 207.

[230] H. Hippler, K. Luther, A. R. Ravishankara and J. Troe, Z. Phys. Chem. 142 (1984) 1.

[231] I. R. Slagle, D. Gutman, J. W. Davies and M. J. Pilling, J. Phys. Chem. 92 (1988) 2455.

[232] D. Walter, H.-H. Grothee, J. W. Davies, M. J. Pilling and A. F. Wagner, Proc. Comb. Inst. 23 (1990) 107.
[233] A. F. Wagner and D. M. Wardlaw, J. Phys. Chem. 92 (1988) 2462.
[234] L. B. Harding and S. J. Klippenstein, work in progress
[235] A. I. Krylov and C. D. Sherrill, J. Chem. Phys. 116 (2002) 3194.
[236] L. V. Slipchenko and A. I. Krylov, J. Chem. Phys. 118 (2003) 6874.
[237] K. Kowalski and P. Piecuch, J. Chem. Phys. 113 (2000) 18.
[238] K. Kowalski and P. Piecuch, Chem. Phys. Lett. 344 (2001) 165.
[239] S. J. Klippenstein and W. D. Allen, Ber. Bunsen. Phys. Chem. 101 (1997) 423.
[240] E. A. Wade, A. Mellinger, M. A. Hall and C. B. Moore, J. Phys. Chem. A 101 (1997) 6568.
[241] S. J. Klippenstein, L. R. Khundkar, A. H. Zewail and R. A. Marcus, J. Chem. Phys. 89 (1988) 4761.
[242] S. J. Klippenstein and R. A. Marcus, J. Chem. Phys. 93 (1990) 2418.
[243] Y. Georgievskii and S. J. Klippenstein, work in progress.
[244] J. M. Bowman, Theor. Chem. Acc. 108 (2002) 125.

Chapter 3

State-specific dynamics of unimolecular dissociation

S. Yu. Grebenshchikov,[a] R. Schinke[a] and W. L. Hase[b]

[a]Max-Planck Institut für Strömungsforschung,
37073 Göttingen, Germany

[b]Department of Chemistry,
Wayne State University, Detroit, Michigan 48202

1. INTRODUCTION

Unimolecular dissociation is one of the simplest types of irreversible chemical reactions: It takes place in a single isolated molecule with an internal energy that exceeds the first dissociation threshold; see Fig. 1(a) for a schematic overview. Nevertheless, the underlying atomic-level reaction mechanisms are very complex. Their theoretical description requires all the power of modern quantum chemical methods, statistical physics and nonlinear dynamics, and even then the full rigor can be achieved just for small, mostly triatomic molecules. Experimental studies have to be likewise advanced: Up to three laser pulses are combined in a modern experiment to elucidate all details of the dissociation process.

The complexity shows up already in the various processes used to excite the reactant molecule above the dissociation threshold. In fact, different excitation mechanisms lead to different dissociation rates. If the molecule is excited through interaction with a buffer gas at statistical equilibrium, which is a common chemical situation [1], the reaction rate $k(T,p)$ is a function of temperature and pressure and, in the high-pressure limit, obeys the well known Arrhenius law. This thermal rate is an average of contributions from many initial states of the excited molecule, which for high pressures are populated according to the Boltzmann distribution. The situation is different in so-called chemical activation experiments, in which the activated molecule is prepared via a bimolecular reaction [2]. This excitation mechanism is more selective, and the energized molecule has a rather well defined internal energy. Chemical activation experiments come close to measuring the monoenergetic rate constant, $k(E)$, which is a function of the excitation energy E. Nevertheless, the uncertainty in the excitation

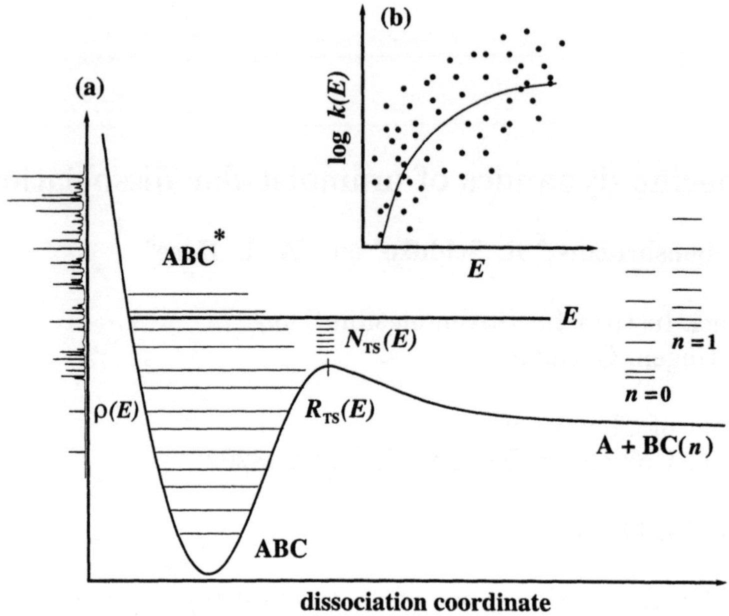

Figure 1: (a) Illustration of the unimolecular dissociation of a reactant molecule ABC into products A and BC. $\rho(E)$, R_{TS} and N_{TS} are the density of reactant states, the intermolecular distance of the transition state (TS) and the number of open channels at the transition state, respectively. The vertical axis on the left-hand side shows a spectrum dominated by sharp resonances. (b) Schematic representation of the energy dependence of the micro-canonical rate constant $k(E)$ (solid line). The dots represent the state-specific quantum mechanical rates k_n.

energy is still significant and $k(E)$ in these experiments has the meaning of an average rate within the excitation bandwidth. The way to reduce this uncertainty is to excite molecules by lasers, as was done in early nano- and picosecond experiments [3–5].

Until the beginning of the 1980's, characterizing unimolecular fragmentation in terms of a micro-canonical rate constant $k(E)$ was the most rigorous level of description practically attainable in experiment and theory. A typical energy dependence of the rate is shown in Fig. 1(b) by the continuous curve. Its characteristic feature is a monotonic increase with rising internal energy, i.e., with increasing number of open reaction channels. The celebrated RRKM model and other versions of the transition state theory [6] were designed to rationalize this behavior and to give recipes for calculating $k(E)$.

The most fundamental description of unimolecular dissociation goes beyond $k(E)$. Although for energies above the dissociation threshold the spectrum of eigenstates of a molecule is, in principal, continuous, quasi-bound (or metastable) states at certain discrete energies E_n can be isolated in the continuum. With modern experimental techniques, using one or more lasers with narrow bandwidths, it is possible to selectively excite these individual eigenstates and to measure the *state-specific* dissociation rates k_n [7]. Both, the micro-canonical and the thermal rate coefficients can be reconstructed by proper averaging of the rates k_n (see, for example, Ref. 6 and the discussion of the Lindeman-Hinshelwood mechanism in Ref. 8).

Already the earliest state-resolved experiments, performed by Moore and coworkers on the dissociation of formaldehyde [9,10], revealed the most striking feature of state-specific rates, i.e., the k_n do not depend smoothly on the excitation energy, but strongly *fluctuate.* The rates of neighboring states in a narrow energy window may differ by several orders of magnitude so that they are best represented as a set of scattered points as illustrated by the dots in Fig. 1(b). The fluctuations are the strongest at the dissociation threshold and diminish gradually as the internal energy increases. However, there is no unique parameter, analogous to temperature, pressure, internal energy or angular momentum, which determines the value of k_n. The systems investigated — experimentally and theoretically — in the past decade beyond any doubt prove that fluctuating rates are characteristic of *all molecules*, those with very small density of states like HCO as well as those with much larger densities of states like H_2CO and NO_2. The fluctuations result from a complicated pattern of energy flow between the reaction coordinate and the rest of the coupled internal degrees of freedom (internal vibrational energy redistribution, IVR [11–13]). This chapter is intended to present the basic principles of the state-specific approach, to survey modern experimental and theoretical methods, and to illustrate their application to several recently studied prototype molecules.

The existence of eigenstates in the dissociation continuum is impressively illustrated in Fig. 2 showing a low-resolution overview spectrum of the protonated formyl radical, DCO — a well studied triatomic with a low density of states [14,15]. The corresponding spectrum for HCO looks similar [16]. Each group of spectral lines, consisting of four individual rotational transitions, represents one vibrational state. Below 5 600 cm^{-1} all lines correspond to true bound states (there are only 29 of them; not all of them are visible in Fig. 2). Above 5 600 cm^{-1} the molecule becomes unbound. Any energy is, in principal, an eigenstate in the continuum so that one would expect the spectrum of DCO to be featureless. Nevertheless, the sequence of spectral lines does not abruptly stop at the dissociation threshold; it persists into the continuum nearly intact, with all lines belonging to the ground electronic state. As high as 1 eV above the threshold the spectrum still consists of well distinguishable peaks. These analogues of the bound

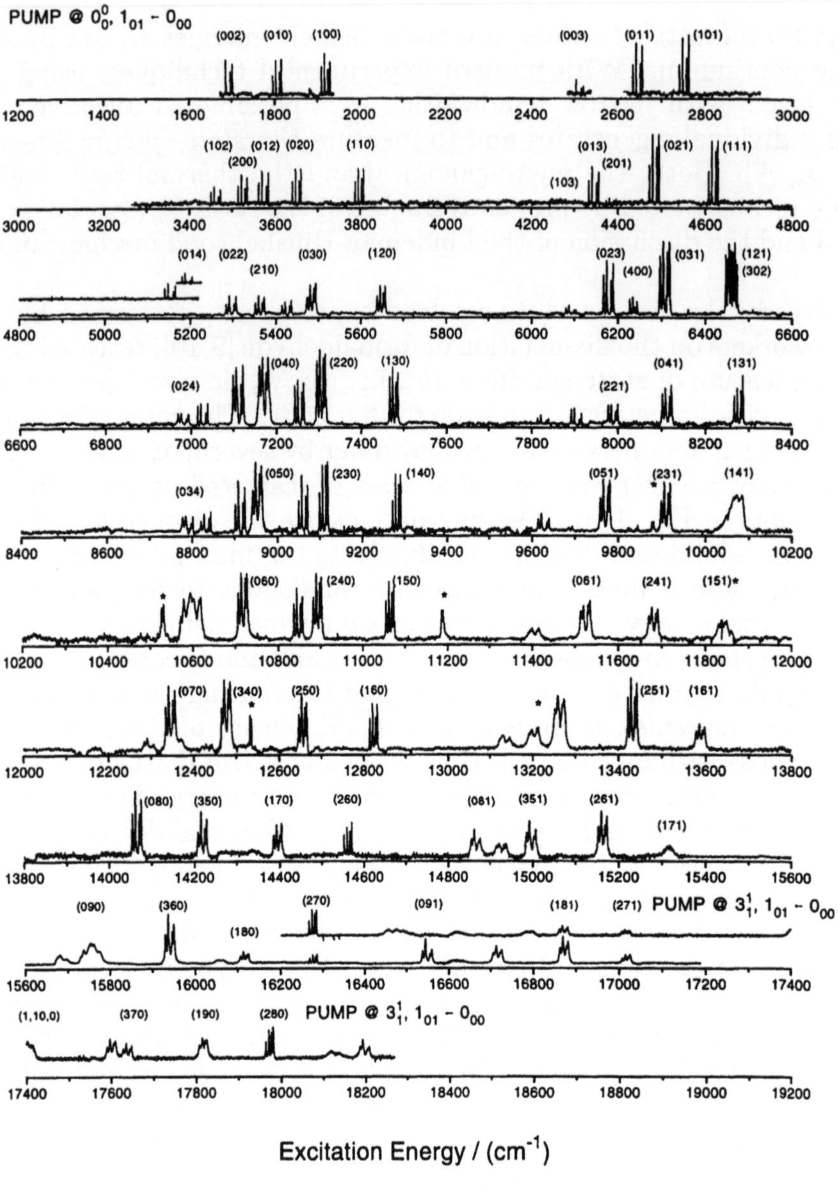

Figure 2: Low-resolution overview stimulated emission pumping spectrum for DCO. The energy is normalized to the ground vibrational state, (0,0,0). Reproduced, with permission of the American Institute of Physics, from Ref. 14.

states above the dissociation threshold are called *resonances* [17–20].

There is, however, a principal difference between bound and resonance states. While the true bound states in Fig. 2 appear as nearly perfect delta-functions, the lines above threshold exhibit homogeneous broadening caused by the finite lifetimes, especially distinct at high excitation energies. According to the energy-time uncertainty relation [18], a width Γ_n in the energy domain implies that the molecule remains in this state only for some time τ_n after which it falls apart. If the line is a Lorentzian, the state decays purely exponentially with a well defined rate constant $k_n = 1/\tau_n = \Gamma_n/\hbar$. Strongly broadened resonance lines in Fig. 2 reflect fast unimolecular decomposition of DCO. As intuition suggests, the decay tends to accelerate as the excitation energy increases; correspondingly, the resonance widths, on average, grow with E. At the same time, the spectral widths do not vary smoothly with E; they fluctuate over an appreciable range. This can be suspected even from the low-resolution spectrum shown in Fig. 2, but it will become much more obvious in the detailed discussion in 5.1. The extent of these fluctuations and their dynamical origins are the central theme of this chapter.

Large fluctuations are known to exist in fluorescence lifetimes of individual rovibrational levels. An intriguing example is the predissociation of $SiH_2(\tilde{A}^1B_1)$ [21]. In this particular case, the dissociation is assumed to include non-Born-Oppenheimer transitions to the $\tilde{X}^1A_1(S_0)$ and the $\tilde{a}^3B_1(T_1)$ state and the irregular variations in the lifetimes are explained in terms of quasi-random variations of the energy spacings between S_0 and T_1 levels. The kind of dissociation processes considered in this chapter are different: They include only a single, namely the ground electronic state.

Frequency-domain spectroscopy is adequate for state-selective studies if the resonances are isolated from one another, i.e., if the average linewidth, $\langle\Gamma\rangle$, is smaller than the average distance between neighboring resonances ($\langle\Gamma\rangle\rho \ll 1$, where ρ is the density of states). As in DCO, one resolves individual resonance levels and thus obtains detailed state-specific information. Strongly fluctuating dissociation rates have been measured for D_2CO [10], HFCO [22], HCO [16], DCO [14], CH_3O [23–25], HOCl [26,27], HN_3 [28] and HONO [29, 30]. In the opposite limit, $\langle\Gamma\rangle\rho \gg 1$, the resonances overlap and can no longer be individually resolved, if there are no selection rules imposed by symmetry. In this case the most unambiguous information about the dissociation rates is provided by time-domain measurements, in which the molecule is prepared in a coherent superposition of a narrow group of eigenstates. The dissociation rate is then extracted from the decay curve of the initial state and represents an average of the state-specific rates k_n contained in the spectral window. The list of molecules, whose unimolecular dissociation has been studied by time-resolved experiments, includes NCNO [31], H_2O_2 [32], CH_2CO [33], NO_2 [34, 35] and CH_3CHO [36]. (See the review of Green et al. [7] for an extensive overview of experiments covering the literature up to 1992.)

Interpretation of state-selective experiments requires exact quantum mechanical calculations. They yield the positions and widths of the resonance states. In quantum dynamics, one solves either the time-dependent or the time-independent Schrödinger equation with the so-called *radiation boundary conditions* [37, 38], which correctly describe the flux of particles leaving the interaction region of the potential and irreversibly falling apart into fragments. Any multi-dimensional quantum calculation requires two equally important ingredients: a global potential energy surface (PES) and an efficient algorithm for locating the eigenvalues of the Hamiltonian.

Experience of the past decade shows that the construction of accurate global PESs, stretching from near-equilibrium geometries to the asymptotic inter-fragment distances, is often the bottleneck of dynamics studies. For this reason full-dimensional quantum calculations for unimolecular dissociation are presently limited to molecules with three or, at best, four atoms, for which the number of internal degrees of freedom and, hence, the numerical effort for creating a global PES are moderate [39, 40]. As a result of significant advances in computational chemistry [41], accurate global PESs have been constructed for a few molecules from ab initio calculations. Figure 3 shows the PES of an important molecule: ozone in its ground electronic state $\tilde{X}^1A'$ [42]. It has a complicated structure with several minima and saddle points. The precise shape of such a PES cannot be surmised without high-level electronic structure calculations. For example, the region of the transition state (TS), i.e., the region where the reactant molecule looses its identity and the fragments are created, is extremely important for the kinetics of the formation and dissociation of ozone [43, 44]. For ozone, the TS is very tight with the lowest frequency at the TS (perpendicular to the reaction coordinate) being of the order of 270 cm^{-1} [45]. The height of the small barrier at the TS depends sensitively on the level of the electronic structure calculations. It is large for a small atomic basis set and gradually decreases with larger and larger basis sizes. If one wants to compare theory with experiment on a high level of detail, very accurate PESs are needed. Potential energy surfaces of other molecules will be discussed below.

Concurrent with the progress in quantum chemistry, new powerful numerical algorithms have been developed for solving the Schrödinger equation for the nuclear motion on these PESs even in the continuum, far away from equilibrium [46]. This and the availability of high-performance computers allowed detailed quantum investigations of resonances for HCO [47–54], DCO [15, 55], HO_2 [56–61], H_3^+ [62], H_3 [63, 64], HNO [39, 65], HOCl [66–69], O_3 [42], H_2O [70, 71], NaHF [72], HeH_2^+ [73], HOCO [74–76], HFCO [77], NO_2 [35] and $ClCH_3Cl^-$ [78]. In some of these cases the theoretical predictions could be directly compared to experimental results on

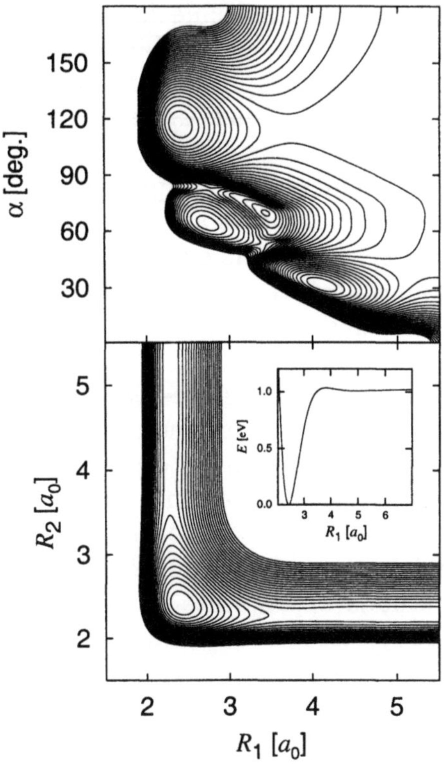

Figure 3: Two-dimensional representations of the ground-state PES of ozone. R_1 and R_2 are the bond distances between the central atom and the two end atoms and α is the OOO bond angle. The inset shows the potential along the minimum energy path. Reproduced, with permission of The Owner Societies, from Ref. 42.

a high level of detail.

Another advantage of the quantum calculations is that they provide a rigorous test of approximate methods for calculating dissociation rates, namely classical trajectories and statistical models. Two commonly used statistical theories are the *Rice-Ramsperger-Kassel-Marcus* (RRKM) theory and the *statistical adiabatic channel model* (SACM). The first one is thoroughly discussed in Chapter 2, while the second one is briefly reviewed in the Introduction. Moreover, the quantum mechanical approach is indispensable in analyzing the reaction mechanisms. A resonance state is characterized not only by its position, width and the distribution of product states, but also by an individual wave function. Analysis of the nodal structure of resonance wave functions gives direct access to the mechanisms of state- and mode-selectivity.

The range of fluctuations of dissociation rates is related to the strength of intra-mode coupling in the molecule [39, 40, 79]. If the coupling is weak and, consequently, the classical dynamics is regular and the quantum states are assignable, the amplitude of fluctuations is large. In this case the dissociation is said to be *mode-specific*. For early theoretical studies of mode-selectivity in molecular dissociation see the work of Manz and coworkers [70,80–82]. With increasing coupling between different degrees of freedom the spectrum gradually loses assignability, the classical dynamics become chaotic and the spread of the decay rates is substantially reduced. The dissociation becomes *statistical state-specific* [83, 84]. However, even in this case the actual range of fluctuations may reach several orders of magnitude!

In this chapter we elucidate the state-specific perspective of unimolecular decomposition of real polyatomic molecules. We will emphasize the quantum mechanical approach and the interpretation of the results of state-of-the-art experiments and calculations in terms of the quantum dynamics of the dissociating molecule. The basis of our discussion is the resonance formulation of unimolecular decay (Sect. 2). Summaries of experimental and numerical methods appropriate for investigating resonances and their decay are the subjects of Sects. 3 and 4, respectively. Sections 5 and 6 are the main parts of the chapter; here, the dissociation rates for several prototype systems are contrasted. In Sect. 5 we shall discuss the mode-specific dissociation of HCO and HOCl, while Sect. 6 concentrates on statistical state-specific dissociation represented by D_2CO and NO_2. Vibrational and rotational product state distributions and the information they carry about the fragmentation step will be discussed in Sect. 7. Our description would be incomplete without alluding to the dissociation dynamics of larger molecules. For them, the only available dynamical method is the use of classical trajectories (Sect. 8). The conclusions and outlook are summarized in Sect. 9.

The dissociation of weakly bound van der Waals complexes is a special case of unimolecular dissociation [20]. Because of the exceedingly weak coupling between the dissociation coordinate and the mode (or modes) initially excited, and the very low density of states of the energized complex, narrow resonances are *the* dominant features of van der Waals spectra. There are, of course, many similarities between the dynamics of *physically bound* and *chemically bound* molecules. The dissociation dynamics of these special molecules (or clusters) has been studied in great detail, both experimentally and theoretically. Exhaustive review articles are available [85–89] and therefore van der Waals molecules will not be particularly considered in this chapter. However, one must keep in mind that, as the density of states of van der Waals molecules increases, their dynamics becomes more and more comparable with the dynamics of strongly bound molecules [90, 91].

Moreover, resonances of the type described in this chapter are also found in bimolecular reactive collisions. For a recent review see Liu, Skodje, and Manolopoulos [92]. They will not be covered here, either.

2. RESONANCE FORMULATION OF UNIMOLECULAR DECAY

Resonances are prominent features in physics and chemistry. They stand out in such diverse fields as nuclear physics (where resonance theory was developed) [93], atomic physics (autoionization of an atom) [94,95], electron-atom scattering [96], photoelectron spectroscopy [97], surface scattering [98], reaction dynamics [92,99], photodissociation [20], van der Waals molecules [100], and scanning tunneling microscopy [101]. Pronounced resonance effects commonly show up also in the frequency dependence of photodissociation (i.e., unimolecular dissociation) cross sections [20, 102]. Traditionally, resonances are considered in the context of scattering theory where they appear in the energy dependence of phase shifts and bimolecular scattering cross sections [17, 103]. Alternatively, resonances can be viewed as (complex) eigenvalues of the Hamilton operator. This formalism was first introduced by Gamow in the problem of α-decay [104], and further developed by Kapur and Peierls [105], Siegert [37], and Schwinger [106] (see also Ref. 107). The eigenvalue approach provides a suitable framework, within which the rates of unimolecular reactions as well as the spectra of dissociating molecules can be rationalized and efficiently calculated. Thorough discussions of resonances as eigenstates can be found in Refs. 38 and 108. In this section we first discuss the principal aspects of this approach using simple one-dimensional models and then point out possible generalizations to the multi-dimensional case.

2.1. One-dimensional square-well model

Since the motion above the dissociation threshold is unbound, the eigenstates can be conveniently represented as linear combinations of spherical waves. The incoming waves, $\exp(-\mathrm{i}kR)/R$, describe particles moving towards the coordinate origin, while the outgoing waves, $\exp(\mathrm{i}kR)/R$, represent motion away from the origin. Here, R is the distance between the dissociation fragments and k is the momentum related to the energy in excess of the dissociation threshold, E, according to $k = \sqrt{2\mu E}/\hbar$; the latter must not be confused with the dissociation rate, which is denoted by the same letter. In scattering, the incoming flux equals the outgoing flux, and the eigenspectrum is real and continuous; any positive kinetic energy is allowed. In dissociation, however, there is no incoming flux — one observes only the flux of fragments which are flying apart. The wave function of a dissociating molecule, ψ, asymptotically contains only an outgoing wave,

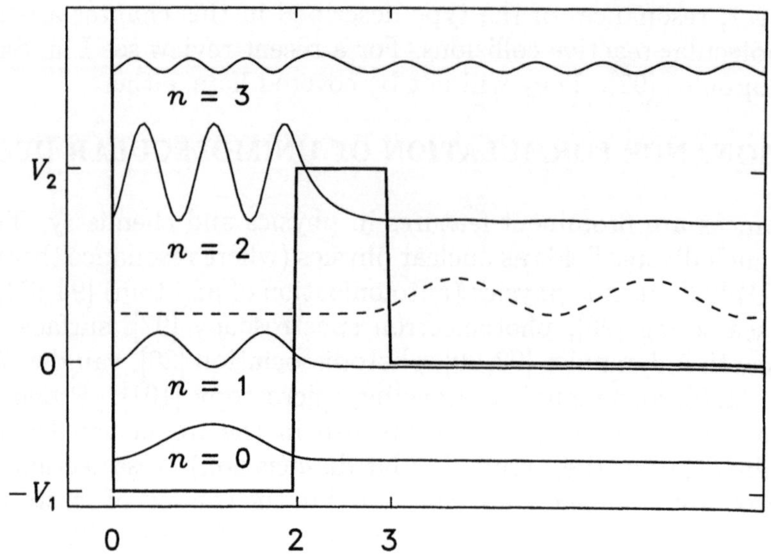

Figure 4: Illustration of resonances in a one-dimensional square-well potential. The two lowest solid lines are bound-state wave functions, whereas the upper two solid lines illustrate resonance wave functions. The dashed curve represents a non-resonant 'scattering state'. Shown is the modulus square of χ; the scaling is different for the different wave functions. In the numerical example in the text, $V_1 = 8$ and $V_2 = 12$.

$\psi \sim \exp(\mathrm{i}kR)/R$. This radiation boundary condition can be satisfied only for a set of distinct eigenvalues and these are the resonance energies.

The concept can be illustrated with a simple one-dimensional problem, in which a spherically symmetric potential well is surrounded by a barrier (see Fig. 4). One might consider this problem as a model for dissociation of a diatomic molecule. According to the general theory, the spectrum is discrete for $E < 0$ and continuous for $E > 0$. Let us now look for unbound solutions, $\chi(R) = R\psi$, of the time-independent Schrödinger equation which behave like $\exp(\mathrm{i}kR)$ at large R. The desired wave functions have the form

$$\begin{aligned} \chi(R) &= A\sin(k_{\mathrm{w}}R) && \text{for } R < 2 && \text{(1a)} \\ &= B\sinh(k_{\mathrm{b}}R + \delta) && \text{for } 2 < R < 3 && \text{(1b)} \\ &= C\exp(\mathrm{i}kR) && \text{for } R > 3, && \text{(1c)} \end{aligned}$$

where $k_{\mathrm{w}} = \sqrt{E + V_1}$ and $k_{\mathrm{b}} = \sqrt{V_2 - E}$ are the wave vectors in the well (w) and at the barrier (b), respectively, and $\hbar = 2\mu = 1$ has been used for convenience. Equation (1b) holds for energies below the barrier; if E

exceeds the barrier $\sinh(x)$ has to be replaced by $\sin(x)$ with $k_b = \sqrt{E - V_2}$. Continuity conditions for the logarithmic derivatives χ'/χ at $R = 2$ and $R = 3$ give a pair of equations,

$$k_\mathrm{b} \coth(2k_\mathrm{b} + \delta) = k_\mathrm{w} \cot(2k_\mathrm{w}) \tag{2a}$$

$$k_\mathrm{b} \coth(3k_\mathrm{b} + \delta) = \mathrm{i}k\,, \tag{2b}$$

analogous to the quantization equations for bound states [18]. Simultaneous solutions of these equations exist only at certain discrete energies E_n. In contrast to the textbook bound-state problem, the eigenvalues E_n are complex numbers, because the radiation boundary condition, Eq. (1c), is also complex.

For the particular example in Fig. 4, the lowest resonance state is found at $E = 8.75 - \mathrm{i}\,0.05$, while the next one has the energy $E = 18.01 - \mathrm{i}\,0.45$. The positive real parts of the eigenenergies, $\mathrm{Re}E \equiv E_0$, indicate that these states are unbound with energies high above the dissociation limit, $E = 0$. The imaginary parts, which will be denoted $-\Gamma/2$, are related to the decay probability in a unit time interval, k [18]. Indeed, the time dependence of the wave function $\chi(R)$ is given by

$$e^{-\mathrm{i}Et/\hbar} = e^{-\mathrm{i}E_0t/\hbar}\, e^{-\Gamma t/2\hbar}\,. \tag{3}$$

The population of particles in the potential well, proportional to $|\chi|^2$, diminishes exponentially, $\sim \exp(-\Gamma t/\hbar)$, so that the decay rate is given by $k = \Gamma/\hbar$. Note, that Eq. (3) explains why physically meaningful resonances lie in the lower half of the complex energy plane: Otherwise, the population would continue to grow in time without bound. For the first resonance, the decay rate is small because the dissociation occurs via tunneling through a high potential barrier. As a result, its lifetime $\tau \sim \hbar/\Gamma$ is long. The states, trapped by a potential barrier, are called shape resonances [103]; in the classical limit, $\hbar \to 0$, the width of a shape resonance vanishes. The energy of the second resonance is significantly larger than the barrier energy and thus trapping behind the barrier is not efficient. Consequently, the rate k and the width Γ are substantial, roughly a factor of ten larger than for the first resonance; its lifetime remains finite even in the classical limit.

Complex eigenenergies, which appear because of the special choice of the boundary condition, are convenient for describing dissociating systems. In experiment, however, one deals exclusively with *real* energies. The physical meaning of the imaginary part of the eigenenergy in a spectroscopic experiment can be understood as follows. The observable in such an experiment is an absorption or emission cross section, $\sigma(E)$. The cross section is proportional to the time-Fourier transform of the autocorrelation function [20], i.e., the overlap between the molecular state (wave packet) immediately after the excitation, $\Phi(t=0)$, and the molecular state at a later time t, $\Phi(t)$. If only one resonance state is initially excited, the shape of $\sigma(E)$ can be shown to behave like [cf. Eq. (3)]

$$\sigma(E) \sim \frac{1}{\hbar} \int e^{-\mathrm{i}E_0 t/\hbar}\, e^{-\Gamma t/2\hbar}\, e^{\mathrm{i}Et/\hbar}\, dt = \frac{(\Gamma/2)}{(E-E_0)^2 + (\Gamma/2)^2} \,. \tag{4}$$

Thus, an exponentially decaying resonance has a Lorentzian line shape. The profile peaks at the real part of the eigenenergy, E_0, while the full width of the spectral line at its half maximum equals Γ. For this reason, resonances are said to have no fixed energy; instead, they are spread over the energy interval $\Delta E \sim \Gamma$ around E_0. What is meant, of course, is that relatively long-lived metastable states can be formed for all energies lying in the interval ΔE. The lifetimes τ of these states obey the same Lorentzian law as the cross section and reach the maximum value, $\hbar/\Gamma$, at the resonance energy E_0.

Further insight into the properties of resonances is provided by the wave functions (Fig. 4). Let us first look at the narrow resonance behind the barrier. The effect of the radiation boundary condition is clearly seen: It selects a state which is mainly concentrated in the potential well. Only the tail of the resonance wave function, with significantly smaller amplitude, extends into the exit channel. Physically, this implies accumulation of particles in the inner region — they spend some time in the well before dissociation takes place. The states which fail to satisfy the outgoing boundary condition have different wave functions. Figure 4 shows an example of a state calculated off resonance (dashed line). Although the energy of this state is lower than that of the resonance, it is localized outside the potential well because of the dominating contribution of the incoming wave. In one dimension, the width grows quickly with energy. The second resonance is already an order of magnitude broader than the first one. Concurrently, the wave function extends far beyond the potential well. The relative amplitude of $|\chi|^2$ in the well is proportional to $1/\Gamma$, so that the narrower the resonance (i.e., the weaker the coupling to the product channel), the higher the probability of finding the particle in the well. Substantial localization in the inner region makes narrow resonance states in molecules accessible through laser excitation in photodissociation experiments. The intensity of a spectral line is proportional to the Franck-Condon factor, $|\langle\chi_0|\chi\rangle|^2$, between an initial bound state χ_0 and the final state χ [20]. The intensity is large only if χ is localized inside the well, and in the continuum of a given electronic state this holds only for the resonances. Scattering states can contribute, at best, only to the background of the absorption spectrum.

Bound-state wave functions possess a distinctive nodal structure, which reflects the assignment of states in terms of vibrational quantum numbers (see the first two bound states in Fig. 4). Often, resonance wave functions, too, can be classified according to the nodal pattern in the well (examples are again provided by Fig. 4). Moreover, on the basis of this assignment both the bound states *and* the resonances can be combined into a single progression. In this sense, *resonances are the continuation of the progression of bound states into the continuum.* The consistency between the eigenstates below and above the dissociation threshold is due to the fact that the boundary conditions for the two types of states resemble each other. The quantization condition for the discrete bound levels suppresses the exponentially growing solution $\exp(+|k|R)$ for $R \to \infty$, while the discrete resonance states result from the requirement that the incoming wave $\exp(-\mathrm{i}kR)$ vanishes asymptotically.

Although the resonance wave functions, defined through Eqs. (1) and (2), provide a physically correct picture in the interaction region of the potential, their practical use requires certain care. The difficulty arises in the asymptotic region. Since the eigenenergy, E_n, has a negative imaginary part, the momentum $\hbar k_n$ also lies in the lower half of the complex plane, $\mathrm{Im}\,\hbar k_n \sim -\Gamma_n$. Therefore, the eigenfunction $\chi_n \sim \exp(\mathrm{i}k_n R)$ exponentially diverges as $R \to \infty$ according to $\exp(+\Gamma k_0 R/4E_0)$. However, if Γ is small compared to the resonance energy E_0, the divergence is slow. Nevertheless, the function χ_n cannot be normalized in the whole configuration space. The normalization integrals are only meaningful if taken over a finite region $R \leq R_{\mathrm{max}}$ (see, for example, Ref. 38). As a result, resonance wave functions possess unusual properties [108–110], and even a special perturbation theory has to be developed for such states [108, 109]. In numerical calculations, their asymptotic divergence must be suppressed through appropriate transformations of the original Hamiltonian, as will be discussed in Sect. 4.

In scattering, resonances are associated with singularities of the **S**-matrix [18], which describes the relative phase of the outgoing wave in the eigenvector at $R \to \infty$:

$$\chi(R \to \infty) = A(E)e^{\mathrm{i}kR} + B(E)e^{-\mathrm{i}kR} \sim S(E)e^{\mathrm{i}kR} + e^{-\mathrm{i}kR}\,. \tag{5}$$

If the amplitude $B(E)$ of the incoming wave in Eq. (5) vanishes, $S(E)$ diverges. Expanding the functions $A(E)$ and $B(E)$ near $E_0 - \mathrm{i}\Gamma/2$, one obtains in first order

$$S(E) = e^{2i\phi_0} \frac{E - (E_0 + i\Gamma/2)}{E - (E_0 - i\Gamma/2)}, \tag{6}$$

that is, resonances correspond to complex simple poles of $S(E)$. The slowly varying phase ϕ_0 describes scattering far away from resonance. Relating the **S**-matrix to the scattering cross section one obtains the famous Breit-Wigner expression [111]:

$$\sigma(E) = \frac{\pi}{k^2} |S(E) - 1|^2 = \frac{\pi}{k^2} \left[\frac{(\Gamma/2)^2}{(E - E_0)^2 + (\Gamma/2)^2} + f_{\text{bg}}(E) + f_{\text{int}}(E) \right]. \tag{7}$$

The first term arises from the resonance scattering, the second term, f_{bg}, is due to the off-resonance phase ϕ_0, while f_{int} describes the interference between the first two. If the background term is small, the cross section reduces to the familiar Lorentzian form and Eq. (7) can be directly used to extract resonance parameters from the experimental or calculated $\sigma(E)$. On the other hand, if f_{bg} cannot be neglected, one encounters a complicated energy dependence of the cross section known as Fano profiles [112].

2.2. Multi-dimensional cases

Up to now we have considered the properties of resonance states in one-dimensional potentials. In the subsequent sections, the discussion will focus on the decay of polyatomic molecules, potential energy surfaces of which depend at least on three variables. Let us now survey — in a more qualitative manner — the new aspects introduced by the additional degrees of freedom. Practical issues of the solution of the Schrödinger equation in more than one dimension are reserved for Sect. 4.

Suppose that a triatomic molecule ABC dissociates into A and BC and let R be the inter-fragment distance between the atom A and the center-of-mass of the diatomic fragment BC. R is the reaction coordinate; two other internal modes are the BC vibration, r, and the ABC bending, γ, which in the course of the reaction gradually turns into free rotation of the diatom. As in one-dimensional problems, the resonance states contain only an outgoing wave, $\exp(ikR)$, at $R \to \infty$. However, the decay occurs now in many product channels simultaneously as the fragments have internal (electronic, vibrational, rotational, etc.) degrees of freedom. Each internal fragment state and an associated set of quantum numbers α define an independent product channel. The distribution of fragments among the final states is unique for each resonance state. Thus, along with E_0 and Γ each resonance

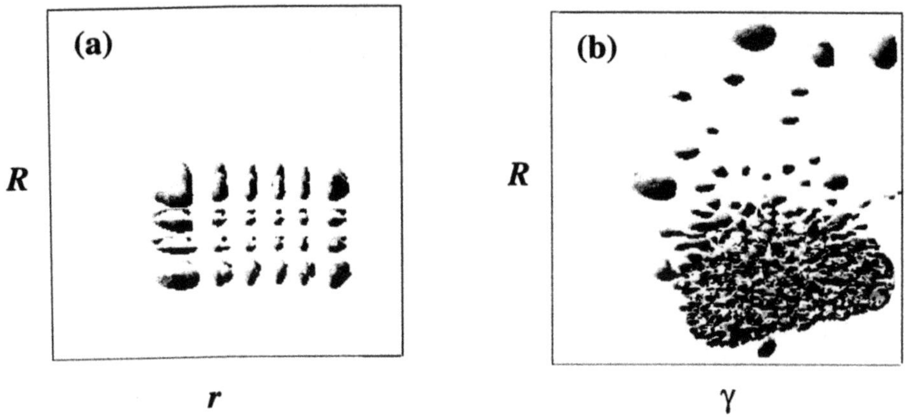

Figure 5: (a) Wave function for resonance state (5,1,3) of HOCl. (b) An example of an unassignable resonance wave function for NO_2. These two molecules will be discussed in more detail in 5.2 and 6.2, respectively.

is characterized by an asymptotic product state distribution (PSD). PSD's can be measured experimentally and provide valuable information on the exit-channel coupling between the fragment degrees of freedom and the dissociation coordinate R [20]. The theoretical description of resonances in multi-channel systems is complicated [17], because the radiation boundary condition must be satisfied in each product channel α [113]. In Sect. 4 we shall discuss several ways to solve (or circumvent) this problem.

Potential energy surfaces of molecules with more than one vibrational coordinate allow for much more complex dynamical patterns than one-dimensional potentials. One can broadly distinguish between regular and chaotic systems [114], which give rise to mode-specific or statistical state-specific dissociation, respectively [83]. Mode specificity implies that the resonance states can be uniquely assigned as many quantum numbers as there are vibrational degrees of freedom. The most direct method to assign a particular state is to look at its wave function. An example of a regular wave function is presented in Fig. 5(a) for the resonance state (5,1,3) of HOCl at an excess energy of $E_0 = 172\,\mathrm{cm}^{-1}$. The width of merely $1.6 \times 10^{-6}\,\mathrm{cm}^{-1}$ is so extremely small, because the energy flow between the HO mode (r) and the dissociation mode (R) for this state is inefficient. In molecules which exhibit mode specificity, both bound and resonance states can be organized in vibrational progressions, in much the same way as it was done for the one-dimensional example above. Resonance widths depend on the quantum numbers of the vibrational modes, and inside each progression they change in a predictable fashion from state to state.

Another illustrative example of mode-specific decay is the dissociation of water. Local-mode and hyperspherical-mode resonance states exist in the same energy range and decay with significantly different rates [70, 80, 81]. Correlations between assignments and vibrational PSD's can also be established [51]. Often, molecules featuring mode-specific decay have low densities of states and shallow potential wells. However, the most important requirement is that the coupling between vibrational modes is weak, so that the dynamics is close to separable. Two examples, HCO and HOCl, will be discussed in Sect. 5.

Statistical state-specific dissociation is typical of molecules in which no consistent assignment of eigenstates is possible. An example for NO_2 is depicted in Fig. 5(b). Although the excess energy is only $35\,\mathrm{cm}^{-1}$, the width of this resonance state is much larger than for the HOCl state illustrated in Fig. 5(a), $0.2\,\mathrm{cm}^{-1}$. Because of the deep well and the three heavy atoms, the de Broglie wavelength in the inner part of the PES is very short in all three coordinates which results in a large number of nodes. Neither clear nodal lines nor vibrational progressions can be isolated and one is restricted to follow the evolution of the resonance widths with the total energy. Irregular nodal patterns are the result of strong coupling between modes which render the classical dynamics of a molecule chaotic. Two examples, H_2CO and NO_2, will be considered in Sect. 6. In such cases, the essential hypothesis of transition state models, that is complete randomization of the energy among all modes, is fulfilled best and it is surmised that the average quantum mechanical rate is well approximated by the RRKM or the SACM rate, whereas the distribution of the rates is predicted by random matrix theory. Finally, there is a broad class of molecules whose behavior is neither mode-specific nor statistical. In these mixed systems a few surviving progressions of resonances are embedded in a 'cloud' of unassignable states.

The most illuminating consequence of multi-dimensional vibrational dynamics in polyatomic molecules are fluctuations of resonance widths. In particular, narrow resonances can often be found far above the first dissociation threshold. We have seen that in systems with one degree of freedom the sequence of resonance states is rather short. Since the excitation energy is deposited directly into the reaction coordinate, the complex breaks apart very quickly and the resonances become broad even close to the dissociation threshold. In polyatomic molecules, energy can be temporarily stored in additional degrees of freedom. The lifetime is then determined not only by the total energy, but also by the rate with which the excitation can be redistributed and transferred to the dissociation bond (see the discussion of the classical phase space structure in Sect. 8).

Fluctuations are inherent to all decaying systems, whether they are regular, irregular, or mixed. Their dynamic origins are most clearly seen in mode-specific dissociation. An excellent example of regular dynamics is provided by hypochlorous acid, HOCl, dissociating into an OH radical and a chlorine atom [26, 27]. Resonance widths in HOCl fluctuate over more than six orders of magnitude [67], the narrowest resonances being pure excitations of the OH mode and the broadest are those in which the dissociation coordinate OH-Cl is strongly excited. For a qualitative discussion, let us first fix the HOCl angle at the equilibrium and consider only the high-frequency OH stretching coordinate r and the low-frequency dissociation coordinate R. In order to understand why extremely narrow resonance states coexist with the broad ones in the same energy range, it is convenient to turn to an adiabatic representation. The vibrationally adiabatic approximation can be applied if $\omega_r/\omega_R \gg 1$ (this frequency ratio is analogous to the well known Massey parameter in scattering). This guarantees that kinematic coupling between the slow (R) and the fast (r) subsystem is weak. In HOCl, $\omega_r/\omega_R \sim 5$ and the adiabatic approximation is expected to be reasonable. In the adiabatic picture, one solves the one-dimensional Schrödinger equation for the fast OH mode keeping the dissociation coordinate fixed. Eigenvalues of the one-dimensional operator are the so-called adiabatic energies $\epsilon_{v_r}(R)$ which are functions of R and also depend parametrically on the OH quantum number v_r. If bending motion is also considered, the adiabatic potentials are potential surfaces, $\epsilon_{v_r}(R,\gamma)$, depending on R and γ. In the second step, the bending motion on these surfaces is quantized yielding one-dimensional potential curves $\epsilon_{v_r,v_\gamma}(R)$. The adiabatic energy curves for $v_r = 0-7$ with $v_\gamma = 0$ are displayed in Fig. 6(a). Each $\epsilon_{v_r,v_\gamma}(R)$ can be further quantized along R, thus giving approximate eigenstates (v_r, v_γ, v_R) of the full 3D problem. The energies of these approximate eigenstates, within a small energy window, are depicted in Fig. 6(b) grouped with respect to the OH stretching quantum number. The density of states in each 'tier' obviously decreases with increasing v_r. The state (7,0,0) lies close to the lowest dissociation threshold. In this energy range, only the states with $v_r = 0$ can adiabatically dissociate. Since for them the dissociation is a one-dimensional process (along the adiabatic curve $v_r = 0$), the resonances with $v_r = 0$ quickly become very broad. All other adiabatic eigenstates with $v_r \neq 0$ are bound because the dissociation thresholds for the adiabats $v_r = 1-7$ lie at higher energies. Therefore, the adiabatic states with $v_r \neq 0$ acquire widths only if vibrationally non-adiabatic transitions are allowed.

Dissociation of a resonance state $(v_r \neq 0, v_\gamma, v_R)$ can be viewed as a multi-step non-adiabatic transition from one adiabatic energy curve to a lower one. Whether single quantum jumps, $\Delta v_r = 1$, or transitions including more than one vibrational quantum, $\Delta v_r > 1$, are more efficient cannot

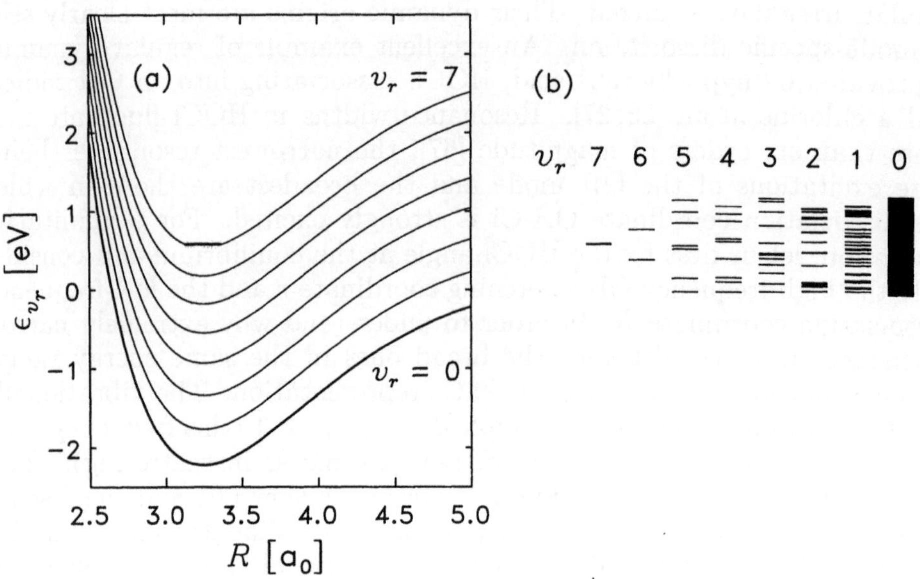

Figure 6: (a) Vibrationally adiabatic potential energy curves $\epsilon_{v_r,v_\gamma}(R)$ with $v_\gamma = 0$ for the dissociation of HOCl. The horizontal line for $v_r = 7$ marks the energy of the overtone state (7,0,0). Only the lowest curve ($v_r = 0$, thick line) is dissociative for energies corresponding to (7,0,0). (b) Approximate energies for the states (v_r, v_γ, v_R) within an energy window of ± 330 cm^{-1} around state (7,0,0), indicated by the shaded box in (a). Note, that the energy scales are different in (a) and (b).

be decided a priori. This depends on the size of the non-adiabatic matrix elements promoting the transitions [20] and on the energy mismatch between the states. Generally speaking, the matrix elements are small if the coupling between the degrees of freedom is weak. For this reason, the widths of resonances with $v_r \neq 0$ are significantly smaller than the widths of the states with $v_r = 0$. The resonances become narrower, as the number of non-adiabatic transitions required for dissociation increases. This mechanism explains the existence of narrow (Feshbach [103]) resonances above the dissociation thresholds of weakly coupled mode-specific systems. Since states belonging to different v_r progressions can be found in the same energy window, the widths fluctuate, and the weaker the coupling, the larger the range of fluctuations.

Irregular quantum systems, which dissociate in a statistical state-specific manner, cannot be analyzed in terms of progressions and the adiabatic picture becomes irrelevant. Nevertheless, the fluctuations have the same physical origin: For each resonance state there is a unique distribution of the total excitation energy among the internal degrees of freedom and, hence, a unique dissociation rate. The main effect of strong intra-mode coupling is to diminish the range of fluctuations relative to the mode-specific systems. At present, rigorous analysis of chaotic systems is possible only through solution of the Schrödinger equation. However, an appealing phenomenological description can be developed within the framework of random matrix theory (for the distribution of widths) combined with transition state models (for the average width) [115–117].

In the preceding discussion we implicitly assumed that the resonance states are isolated from one another and the separation between them, ΔE, is larger than their average width, $\langle\Gamma\rangle$. However, since both the density of states, $\rho \sim 1/\Delta E$, and the average width grow with energy, it is inevitable that at some E the parameter

$$\xi = \langle\Gamma\rangle/\Delta E = \langle\Gamma\rangle\rho \tag{8}$$

will become larger than unity and the resonances will start to overlap. Due to interference between adjacent states in the overlapping regime, the absorption spectra consist of irregularly shaped lines (Ericson fluctuations [118]), which have little in common with the Lorentzians of the isolated resonances. The theory of overlapping resonances in unimolecular dissociation, pioneered by Mies and Krauss [119], is still under development (see, for example, Refs. 120 and 121).

2.3. Mixing between resonances: a simple model

A common feature in molecules is the mixing between states, which may strongly affect the spectroscopic constants [122] and, of course, also the resonance width. The influence of this mixing can be elucidated by introducing a parameter to the Hamiltonian and watching how the width changes with this parameter.

The one-dimensional potential depicted in Fig. 7(a) provides an illustration of this effect. The Schrödinger equation can be solved with the method used for the square-well case above. Each well gives rise to a nearly independent progression of states. For $\hbar = 2\mu = 1$ and other potential parameters indicated in Fig. 7 one finds that the system has two bound states and a resonance state at $9.46 - \mathrm{i}\,0.11$ localized above the deep outer well. There is also another resonance in the system, $E = 9.8 - \mathrm{i}\,0.002$. Its width is very small because this state belongs to the shallow inner well, which is separated from the continuum by a potential barrier. Suppose that we force — by varying a parameter in the Hamiltonian — the narrow state (denoted **n**) in the shallow well to move across the broader resonance (**b**) belonging to the deep minimum. The relative positions of the two states can be, for example, controlled by shifting the infinite wall at the

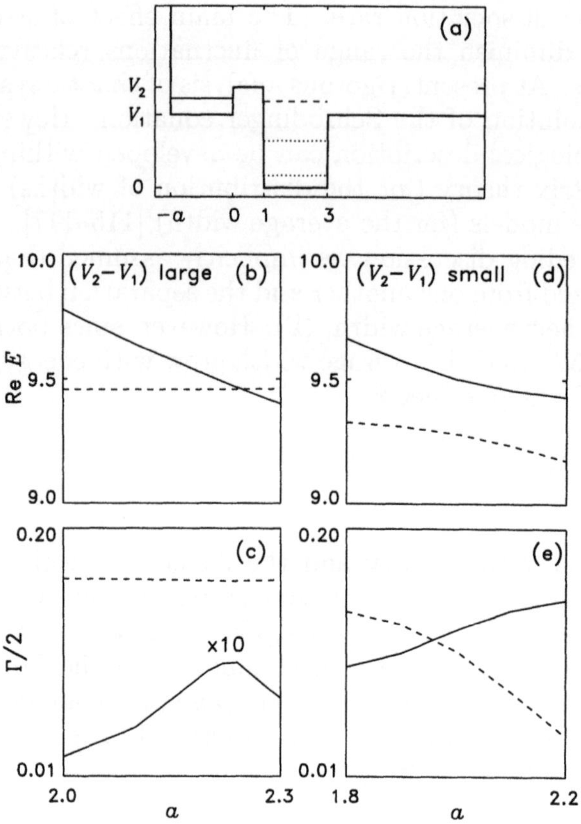

Figure 7: Illustration of the mixing between two resonances — for the double-well potential shown in (a) — as a function of the potential parameter a (the width of the inner potential well) and its influence on the resonance energies (middle two panels) and widths Γ (lower two panels, plotted on logarithmic scales). Solid lines represent the narrow resonance **n** located in the inner well, while the dashed lines indicate the broad resonance **b** localized in the outer well. The weak-coupling limit is shown in (b) and (c), while the strong-coupling limit is illustrated in (d) and (e). In the example discussed in the text, $V_1 = 8$ and $V_2 = 17$ for (b) and (c) and $V_2 = 11$ for (d) and (e).

left and thus changing the distance a. As a increases and the well widens, **n** shifts to a lower energy. The actual change of the energies and widths of the states upon slow variation of the control parameter a depends on the strength of coupling (i.e., the height $V_2 - V_1$ of the potential barrier) between the two wells. If the coupling is weak [Fig. 7(b) and (c)], the real part of the eigenenergies simply cross each other. This does not contradict Wigner's non-crossing rule [18] because the true eigenstates in our case are complex. At the same time, the widths of the resonances show an avoided crossing. The avoided crossing character becomes clearer when the coupling between the two wells is larger. The difference between Γ_n and Γ_b is nearly two orders of magnitude (logarithmic scale!). The broader state remains almost unaffected by the parameter variation, while the width Γ_n changes by a factor of four over the crossing region.

Completely different is the pattern of resonance interaction in the strong-coupling limit, when the potential barrier is low [Fig. 7(d) and (e)]. Now, the avoided crossing is observed for the real parts of the eigenenergies, while the two widths, which are of the same order of magnitude, exhibit a crossing. The resonance wave functions remain localized in the respective wells in the weak-coupling case, but they completely change their individuality upon crossing if the coupling is strong. The crossing patterns observed in Fig. 7 can be easily explained using perturbation theory for a two-level system with complex eigenvalues [123].

Mixing effects as illustrated by this ultra simple two-well one-dimensional system are observed also in more complicated systems, for example, resonances in tunable microwave cavities [123,124] and in quantum billiards connected to a waveguide [125]. Examples from the molecular world are provided by the state-specific decay of triatomic molecules like HOCl [126], HCO [127], and HNO [128]. They will be discussed in more detail in 5.3. Note, that in a molecule with several degrees of freedom and a high density of states, many resonances are simultaneously involved in mutual interactions. Understanding the impact of these dynamical interactions on the overall distribution of the decay rates remains a challenging problem.

3. EXPERIMENTAL APPROACHES

Two basic problems have to be overcome in order to measure state-resolved dissociation rates: First, the resonance states have to be accessed and, second, the resolution must be sufficiently high to guarantee that a single rotational-vibrational state is excited rather than a coherent superposition of several states. The latter of these problems has been circumvented by using various combinations of cooling in a supersonic free-jet, high-resolution (i.e., single-mode) lasers, and double-resonance methods (see below). However, the technical requirements for obtaining full state resolution become more and more severe as the density of states increases. Therefore, even with the most advanced techniques currently available state-resolved rates can be measured only for small molecules, triatoms or molecules with four atoms, of which at least one is a hydrogen atom. The preparation of the reactant molecule at energies at or above threshold in most experiments is achieved by one of three generic ways, which will be illustrated in this section: promotion to an excited electronic state followed by an electronically non-adiabatic transition to the ground state and subsequent dissociation on the ground-state PES, overtone pumping, and stimulated emission pumping. This section is not intended to provide an exhaustive overview of the various experimental techniques; merely the basic schemes will be briefly outlined — from a theorist's perspective. For more details see Refs. 6, 7, and 126 and references therein.

3.1. Preparation by electronic excitation

In many molecules an excited electronic state is embedded in the potential well of the ground electronic state. NO_2 [129] (Fig. 8) and H_2CO are prototypes. In H_2CO, for example, the lowest vibrational state of the S_1 electronic state is close to the dissociation barrier of the S_0 electronic state [10]. A $S_1 \leftarrow S_0$ transition thus prepares the molecule at energies in the region of the first dissociation threshold of S_0. If the excited state is bound in this energy range, the molecule can dissociate only via a non-radiative transition (*internal conversion* or *intersystem crossing*) back to

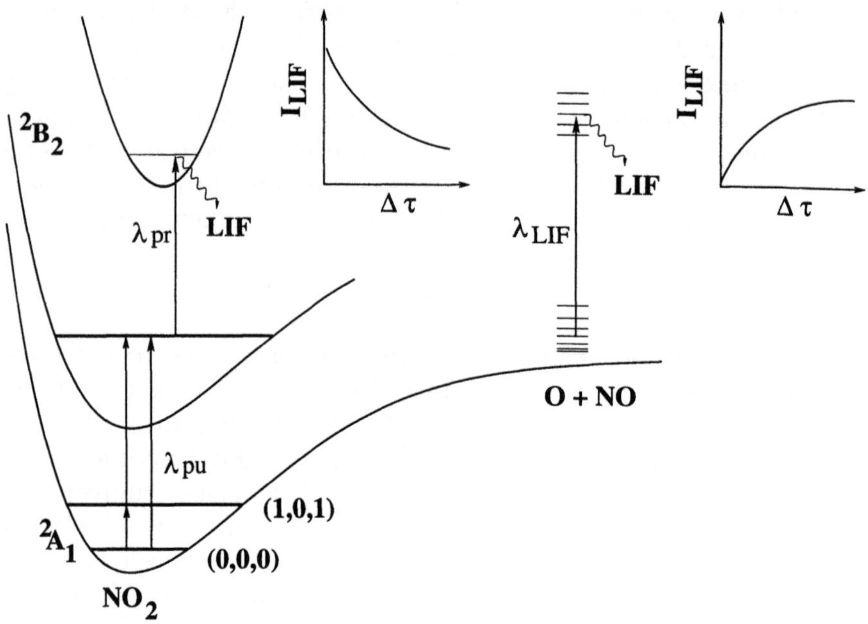

Figure 8: Illustration of the various excitation schemes used to study the unimolecular dissociation of NO_2. The dissociation laser with wavelength λ_{pu} excites strongly mixed $^2A_1/^2B_2$ states near the dissociation threshold. The population of excited NO_2 molecules is monitored by excitation to a higher electronic state with a second laser (λ_{pr}) and collecting the laser-induced-fluorescence (LIF) intensity, I_{LIF}. The decay of I_{LIF} as a function of the delay time, $\Delta\tau$, between pump and probe laser provides direct information about the dissociation of highly excited NO_2. Alternatively, the appearance of the products can be monitored by tuning a specific transition of NO, λ_{LIF}, and detecting the LIF signal.

the ground state where the highly energized molecule subsequently breaks apart. The non-adiabatic coupling can be weak like in H_2CO, or it can be very strong as in NO_2 [130].

The following discussion is centered around NO_2 and illustrated in Fig. 8. Since an electronic transition with a large change in the nuclear geometry (which implies appreciable Franck-Condon factors) is involved, the highly excited states can be accessed by one-photon absorption (λ_{pu}). The oxygen atom [131] or the NO radical [132, 133] can be probed by laser-induced-fluorescence (LIF, λ_{LIF}). Measuring the LIF intensity as function of the excitation wavelength λ_{pu} yields a so-called *photofragment excitation* (PHOFEX) or *photofragment yield* (PHOFRY) *spectrum.* Such spectra show resonance structures in the same way as the absorption or emission spectrum, even more resolved because only one single product channel is considered [20]. In order to minimize congestion due to the many initially populated rotational states of the reactant, it is necessary to work at very low temperatures, in the range of a few K or so; in the experiment of Miyawaki et al. [134] a rotational temperature as low as 0.9 K was achieved. Alternatively, one can use *double-resonance* (DR) *spectroscopy* for minimizing spectral congestion. There are several possibilities of DR spectroscopy. Reid et al. [135, 136] used an IR-visible excitation scheme, where specific rotational states in the $(1,0,1) \leftarrow (0,0,0)$ band of NO_2 are first excited with tunable infrared radiation (cf. Fig. 8). A second laser promotes the pre-excited molecules into the continuum. Abel et al. [137] used a different type of DR spectroscopy to record a spectrum just a few cm^{-1} above threshold, that is *fluorescence dip spectroscopy.* One laser is tuned to a transition between a particular vibrational-rotational state in the ground electronic state and a particular level in a long-lived excited electronic state, for example $2\,^2B_2$ (not shown in Fig. 8). The LIF signal from this state is the observable quantity. The wavelength of a second laser, the dissociation laser with λ_{pu}, is scanned and induces transitions in the heavily mixed $^2A_1/^2B_2 \leftarrow \tilde{X}^2A_1$ electronic system. The first laser is weak whereas the second one is strong. If the two transitions share one common level, for example the ground state level (0,0,0), the dissociation laser creates a population hole in this level and thereby decreases the fluorescence from the excited electronic state, induced by the first laser. An example of such a DR spectrum is shown in Fig. 9. The individual lines indicate the same kind of resonances as seen for DCO in Fig. 2. Estimates of line widths have been obtained, however, only in an exceedingly narrow $10\,cm^{-1}$ energy interval above threshold. Because of the relatively large density of states, the individual resonances rapidly begin to overlap with increasing energy and this makes the determination of meaningful widths impossible.

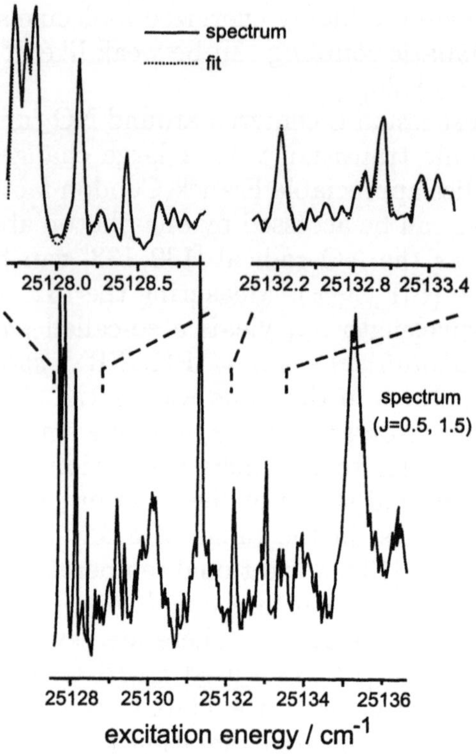

Figure 9: The lower trace shows the experimental double-resonance spectrum of NO_2 just above the dissociation threshold. The upper traces show enlargements including simulations (dotted lines) of the spectrum with multi-line fits (Lorentzian profiles). Courtesy of B. Abel.

If congestion or severe overlap forbids the extraction of rates, information about the decay time can be obtained only via time-resolved measurements [138–140]. In a first step, the reactant molecule is excited with a short pulse, whose spectral width is broad compared to the level spacing. In the second step, a second laser (λ_{pr}), which is delayed with respect to the first one by $\Delta\tau$, promotes the highly excited molecules to a higher electronic state [35]. This state must live for a sufficiently long time in order to allow the excited molecule to fluoresce. The LIF signal, which is a measure of the number of excited NO_2 molecules still intact at time $\Delta\tau$, then decreases as function of $\Delta\tau$ as illustrated in Fig. 8. If the decay curve is nearly exponential, one can define a dissociation rate constant $k(E)$. However, this rate is averaged over all the eigenstates that are coherently excited by the first pulse. Alternatively, one can monitor the appearance of the fragments by probing one particular transition of NO and monitoring the LIF signal from the fragments as a function of $\Delta\tau$ [34]. Decay curves for NO_2 will be discussed in 6.2.

The list of molecules, which are excited above the dissociation threshold by first pumping an excited electronic state which is non-adiabatically coupled to the ground state, is long [7]. Molecules discussed in this chapter are nitrogen dioxide, formaldehyde and ketene.

3.2. Overtone pumping

Another method of exciting a molecule to very high vibrational levels is overtone excitation: A particular mode is directly excited by several quanta [141, 142]. This scheme requires, however, that the intramolecular motion in this mode is robust up to high energies, i.e., that the corresponding wave functions are not much perturbed through mixing with nearby states. The vibrational motion of a stretching vibration including a light atom (e.g., hydrogen) often fulfills this requirement. The corresponding frequency is large compared to the other frequencies giving rise to an adiabatic decoupling scheme as described in 2.2 (Fig. 6). HOCl, which will be discussed in detail in 5.2, is an illustrative example.

A typical overtone pumping experiment, with double-resonance excitation, is illustrated in the left-hand panel of Fig. 10 [126]. It involves three different laser pulses of 5–8 ns duration. The first laser promotes molecules to a specific vibrational-rotational intermediate state with two quanta in the OH stretch vibration. After a delay time the second pulse promotes only the preselected molecules to a higher OH stretch overtone level, that lies in the continuum and dissociates to OH and Cl. Finally, the OH fragments are detected by LIF using a third laser.

Scanning the frequency of the dissociation laser and collecting the total OH fluorescence, while the state-selection and probe frequencies are kept fixed on specific transitions, produces a PHOFEX spectrum; an example is displayed in the right-hand panel of Fig. 10. The lines correspond to specific resonance states with rotational quantum number J and projection quantum number $K = 2$ in vibrational state (6,0,0). If the individual lines are broader than the resolution of the laser system, one can determine the width from fitting the spectrum and thus determine the state-specific dissociation rate. If the true linewidth caused by dissociation is smaller than the resolution of the laser system, the rates can be extracted from time-resolved measurements. All three laser frequencies are fixed, and the OH probe laser used to detect a particular state of OH is delayed with respect to the dissociation laser. In this way one can monitor the appearance of the OH products as function of the delay time, in the same way as described above for NO_2. In contrast to NO_2, however, the rate is a state-specific rate rather than an average rate, because of the high selectivity of the overtone

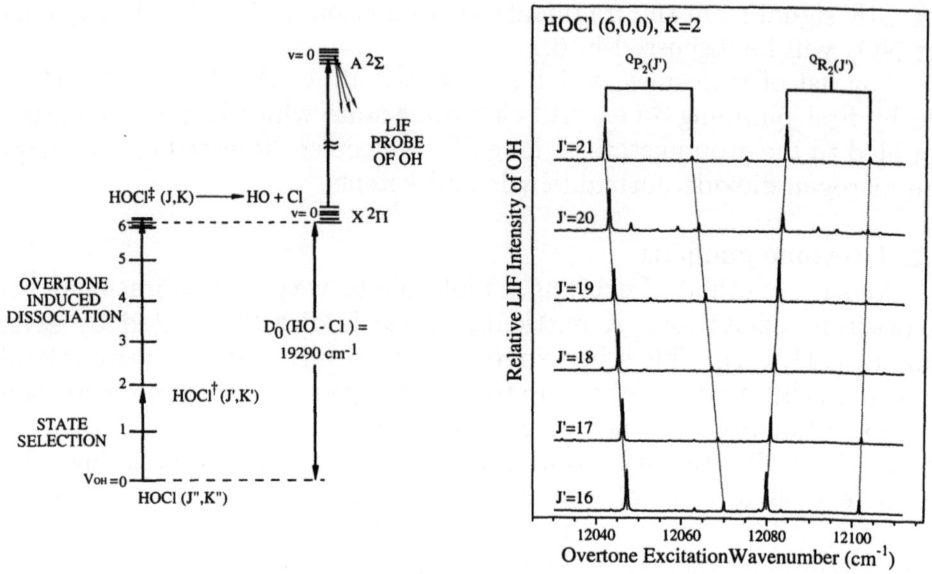

Figure 10: The left-hand panel shows the excitation scheme used in a typical double-resonance overtone photofragment spectroscopy experiment. The particular scheme shown here is used to prepare HOCl molecules with six quanta of OH stretching and many quanta of rotation, just above the dissociation threshold. The right-hand panel depicts sample double-resonance spectra of the HOCl($6v_1 \leftarrow 2v_1$) band. Each spectrum in the plot originates from a selected (J', K') rotational state. Courtesy of T. R. Rizzo.

pumping experiment. The advantage of the double-resonance technique is that cooling is not essential. Therefore it is possible to study the variation of the rate with the rotational quantum numbers J and K over the entire range populated according to the Boltzmann distribution at room temperature. This has been an important aspect in the dissociation of HOCl and will be elucidated below in 5.2. The unimolecular dissociations of hydrogen peroxide, HOOH, [143–146] and HN_3 [28] are other examples that have been studied by overtone pumping.

3.3. Stimulated emission pumping

Overtone pumping spectroscopy has the limitation that, because excitation starts in the ground vibrational state, only very specific states can be accessed, e.g., the $(v_1, 0, 0)$ progression in HOCl and only few other states in the vicinity of the $(v_1, 0, 0)$ states. Stimulated emission pumping (SEP) [147], on the other hand, involves a transition to an excited electronic state, whose equilibrium geometry may be quite different from the equilibrium in the ground state. Therefore, Franck-Condon factors are comparatively large for a wide variety of vibrational states, not just the

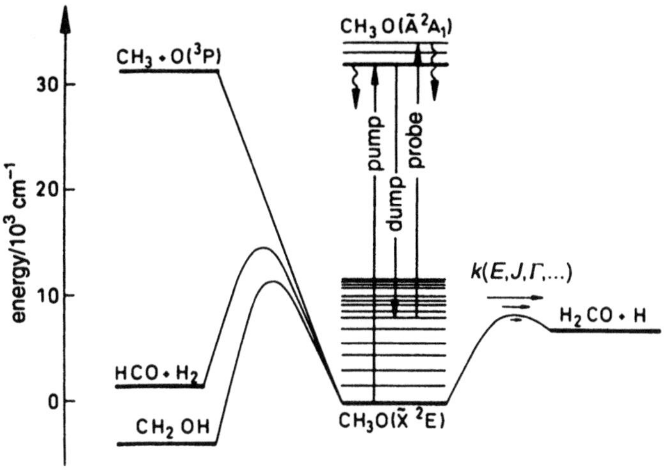

Figure 11: Schematic illustration of an SEP spectroscopy experiment. Reprinted, with permission of the Royal Society of Chemistry, from Ref. 25.

overtones of one particular mode.

Figure 11 illustrates the principal scheme for CH_3O [25]. The pump laser promotes the molecule to a specific rovibrational level in a long-lived excited electronic state and the fluorescence is monitored. The dump laser, which is delayed relative to the pump pulse, leads to a depletion of the initially excited state, provided its frequency matches a transition to a specific level in the ground-state manifold. Thus, scanning the frequency of the dump laser yields a fluorescence dip spectrum as shown in Fig. 12 for CH_3O. The DCO spectrum shown in Fig. 2 has also been recorded in this way. The widths of the resonance peaks depend dramatically on the total energy; in the highest energy window shown in Fig. 12 the resonances strongly overlap so that instead of single lines the spectrum exhibits broad diffuse structures. The non-monotonic dependence of Γ on energy should be noted.

If the lifetime of the excited resonance state is too long for direct measurement of the rate via the widths of the spectral features, one can use a third laser (the probe laser in Fig. 11) to resonantly promote the molecules from this level to a rovibrational level in the excited electronic state. The decrease of the total LIF signal as function of the delay time between pump and probe laser reflects the state-specific dissociation rate. The limitation of the SEP technique is that an excited state has to be found, which lives long enough and which is accessible by all three lasers. Molecules, which have been studied by SEP spectroscopy in the context of unimolecular dissociation, are HCO, DCO, HFCO and CH_3O.

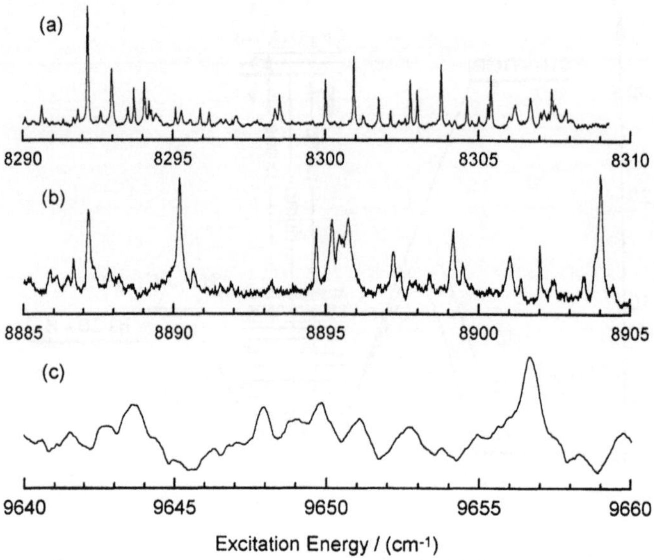

Figure 12: High-resolution SEP spectrum of CH_3O. Reprinted, with permission of the Royal Society of Chemistry, from Ref. 25.

4. COMPUTATIONAL METHODS

In this section, we focus on practical aspects of computing resonance spectra in polyatomic molecules. The main problem is to solve the Schrödinger equation for the nuclear motion in its time-independent,

$$\hat{H}\Psi = E\Psi \,, \tag{9}$$

or time-dependent form,

$$i\hbar\frac{\partial\Phi}{\partial t} = \hat{H}\Phi \,. \tag{10}$$

We shall first outline the major computational strategies and then select two different methods for a more detailed presentation.

In order to keep the presentation as simple as possible, we will limit the discussion to triatomic molecules and motion on a single PES (Born-Oppenheimer approximation). Moreover, only one chemical arrangement channel is considered in the reaction ABC → A + BC. Formal extensions to more general situations are straightforward. The dynamics, then, can be treated in a single set of Jacobi coordinates (R, r, γ). Here, R is the reaction coordinate, i.e., the distance between atom A and the center-of-mass of BC, r is the vibrational coordinate of BC, and γ is the angle between the vectors $\mathbf{R}$ and $\mathbf{r}$. For total angular momentum $J = 0$, the Hamiltonian has the well known form [20]

$$\hat{H} = -\frac{\hbar^2}{2\mu_R}\frac{\partial^2}{\partial R^2} - \frac{\hbar^2}{2\mu_r}\frac{\partial^2}{\partial r^2} + B(R,r)\hat{j}^2 + V(R,r,\gamma)\,, \tag{11}$$

where $\hat{j}$ is the rotational angular momentum operator of BC, $B(R,r) = \hbar^2(2\mu_R R^2)^{-1} + \hbar^2(2\mu_r r^2)^{-1}$, and μ_R and μ_r are the reduced masses of A-BC and BC, respectively. The Hamiltonian for rotating molecules can be found, for example, in Ref. 20. The general numerical schemes for $J \neq 0$ are the same as for rotationless molecules; the notation and implementation, however, become more involved.

4.1. Overview

Numerical methods for calculating resonance parameters can be broadly classified into two groups. The positions and widths can be obtained as complex eigenvalues of the Hamiltonian $\hat{H}$ with eigenfunctions satisfying the radiation boundary conditions. This is a so-called *direct* approach, which requires diagonalization of a complex symmetric Hamiltonian matrix. Alternatively, one can work exclusively on the real energy axis and construct the continuum (scattering) wave functions. These wave functions are further used to calculate observables which exhibit resonance behavior as functions of energy. The desired positions and widths are extracted from this energy dependence. Such an approach is referred to as *indirect*. It shares many common features with computational schemes developed for inelastic or reactive scattering [46]. Both direct and indirect approaches have advantages as well as limitations and the choice of an optimal method depends on the problem. However, in the past few years direct matrix methods, for which some very efficient implementations have been developed, have become more and more popular. In the rest of this subsection we sketch the most important numerical techniques for investigating resonances. Extended discussions of one particular indirect and one particular direct method are given in 4.2 and 4.3, respectively.

4.1.1. Indirect approaches

The standard indirect approach for calculations in the continuum is to solve the close coupling equations [20, 148]. The first investigation of resonance states for HCO [149, 150] and the calculations of Kendrick and Pack [56] for HO_2 have been performed using this universal method. In this approach, widely used in scattering, the continuum wave function at a given energy E is represented in the basis of the fragment eigenstates, so that the time-independent Schrödinger equation, Eq. (9), is transformed into a set of coupled ordinary differential equations along the reaction coordinate R. N linearly independent wave functions, each of which is forced to be zero as $R \to 0$ (regular solutions) are constructed by integrating the system of N differential equations at energy E from small R distances to the asymptotic region. With these functions proper state wave functions $\Psi^{(\alpha)}$ are constructed, which obey the known asymptotic scattering boundary conditions in each open channel α. Determining such states at a number of energies allows one to infer the resonance parameters. In doing so, one can, for example, calculate the trace of the Wigner-Smith time-delay matrix [151, 152],

$$\mathbf{Q}(E) = i\hbar \mathbf{S}(E) \frac{d\mathbf{S}^{\dagger}(E)}{dE} . \tag{12}$$

Since resonances correspond to poles of the **S**-matrix (see 2.1), $\mathrm{Tr}\, \mathbf{Q}(E)$ has a familiar Lorentzian shape in the vicinity of each isolated resonance. The positions and widths can be determined from a non-linear fit to the Breit-Wigner form, Eq. (7) [40]. Another option was chosen by Dobbyn et al. in studies of the dissociation of the HO_2 radical [60]. They overlapped the scattering state $\Psi^{(\alpha)}$ in each open channel α with some arbitrary wave packet Ψ_0, localized in the interaction region of the potential, and constructed an artificial photo-absorption spectrum $\sigma(E)$, which is a sum of partial contributions $\sigma_\alpha(E)$ [20], i.e.,

$$\sigma(E) = \sum_{\alpha} \sigma_\alpha(E) \sim \sum_{\alpha} |\langle \Psi_0 | \Psi^{(\alpha)} \rangle|^2 . \tag{13}$$

The resulting cross section $\sigma(E)$ consists of Lorentzians which correspond to resonance states. A calculated spectrum for DCO, constructed in this way, will be shown below.

The major advantage of the close coupling method is that the scattering boundary conditions are *exactly* satisfied in all open fragment channels. On the other hand, the whole calculation is performed for each scattering energy separately, and the estimation of the energy-dependent lifetime matrix or the cross section can become prohibitively time consuming, especially if the resonances are narrow and the energy grid has to be dense. A new generation of indirect methods, based on the representation of the Hamiltonian in a basis of square integrable (L^2) functions, effectively solves the problem of scanning the scattering energy [53, 60, 153–160]. As in the close coupling approach, a proper scattering wave is constructed from a set of linearly independent regular solutions. However, the method of calculating these regular solutions is different. They are obtained by first diagonalizing the Hamiltonian in an L^2 basis in the inner part of the interaction region and then matching the resulting functions with the analytical asymptotic solution. Typically, diagonalization, performed only once, is the expensive part of the calculation, while the matching procedure for each scattering energy is very fast. Linearly independent waves can be generated within the framework of the time-independent wave packet approach [153–155] or the artificial boundary inhomogeneity method [53, 156]. Another scheme, in which the L^2 diagonalization and the subsequent matching are effectively combined to give the global wave functions and the **S**-matrix, is based on the log-derivative version of the Kohn variational principle [60, 157–159, 161] and is closely related to the **R**-matrix theory of Wigner and Eisenbud [162]. One particular implementation of this scheme, used in calculations of resonance spectra of HO_2 (Ref. 60) and HCO (Ref. 51), is discussed in 4.2.

Another indirect technique, which enables one to calculate the resonance spectrum for many energies simultaneously, is the wave packet method [163–168], used, for example, in studies of non-rotating [47, 169] and rotating [52] HCO, HO_2 [170] and H_2O [71]. One iteratively solves the time-dependent Schrödinger equation, Eq. (10),

$$\Phi(t+\tau) = e^{-\mathrm{i}\hat{H}\tau/\hbar}\,\Phi(t)\,, \tag{14}$$

starting with a well defined wave packet $\Phi(0)$ localized in the interaction part of the potential. Central to the method is the relation between the absorption spectrum of the system and the autocorrelation function $S(t) = \langle\Phi(0)|\Phi(t)\rangle$, i.e., the overlap between the wave packet at $t = 0$ and at $t > 0$ (see, for example, Ref. 20),

$$\sigma(E) \sim \int_0^\infty S(t)\,e^{\mathrm{i}Et/\hbar}\,dt\,. \tag{15}$$

As in the case of the artificial photo-absorption spectrum above, the spectrum $\sigma(E)$ features characteristic resonance profiles from which the positions and widths of all states, represented in the initial wave packet, can be extracted. If $\Phi(0)$ overlaps with all eigenstates in a broad energy range, the entire spectrum can be recovered in a single calculation. In practical applications, the exponential operator (the propagator) in Eq. (14) is evaluated as a polynomial series in $\hat{H}$. The expansion can be performed with Chebyshev [171] or other orthogonal polynomials [52, 172], using the Lanczos iterative scheme [169,173], or a split-operator method [174]. In all realizations, the core of the procedure is the multiplication of the Hamiltonian with the state vector Φ, and there exist highly efficient algorithms for this essential step, which employ the Fast Fourier Transform (FFT) method [175] or the Discrete Variable Representation (DVR) [176].

The propagation of wave packets has several advantages — it is easy to implement, global in energy, and easy to apply to several arrangement channels. It has, however, also several significant limitations. First, it is mainly designed to produce the spectrum, not the resonance wave functions and these can be obtained only at the extra cost of repeating the whole calculation again and Fourier transforming the wave packet at the resonance energies 'on the fly' [20]. Second, the proper boundary conditions are difficult to apply to the part of the wave packet which has left the interaction zone. Only a finite portion of the configuration space can be considered in the calculation, and the wave packet has to be suppressed in the asymptotic region in order to avoid undesired reflections at the edge of the numerical grid. This approximate technique, known as the absorbing potential method [164, 177], amounts to augmenting the original molecular Hamiltonian with a small energy-dependent imaginary part. Of course, the absorption is seldom perfect and a part of the wave packet is reflected, thereby corrupting the autocorrelation function and affecting the peaks in the spectrum. The shorter the propagation time, the less sensitive are the calculations to the boundary of the configuration space. If $\sigma(E)$ is extracted from the autocorrelation function via Fourier transformation, as in Eq. (15), the energy resolution, δE, is related to the propagation time T through the energy-time uncertainty principle, $\delta E\, T \sim \hbar$. Thus, the closely spaced spectral peaks are resolved only after a long propagation. An actual calculation must be carefully balanced before a resolved spectrum is obtained with the desired accuracy.

Restrictions on the time resolution, imposed by the uncertainty principle, are noticeably relaxed if the autocorrelation function is analyzed without using the Fourier transform. Various signal processing algorithms, for example Prony's method [178, 179], ESPRIT [180], or MUSIC [179] allow one to extract the spectral content from shorter time signals. Recently, another processing scheme, called low-storage filter diagonalization method, was proposed for quantum signals calculated on an evenly spaced time grid [181–185]. Using it, one can effectively solve the harmonic inversion problem and perform a non-linear fit of the signal to a finite sum of exponential functions $\exp(\mathrm{i}\omega_n t)$ with complex frequencies ω_n. The idea is to reconstruct the matrix elements of an effective time-evolution operator, which governs the quantum dynamics in a given small energy range, directly from the autocorrelation function and then to find the eigenvalues of the operator using standard diagonalization techniques. The method was successfully used in calculating the resonance spectrum of such notoriously complicated molecules as H_3^+ (Ref. 182) and NO_2 (Ref. 35). Furthermore, another version of this approach, the regularized resolvent transform [186], can be used to estimate the resolvent of the evolution operator directly, without first analyzing the autocorrelation function.

In indirect methods, the resonance parameters are determined from the energy dependence of the absorption spectrum. An important extra step — the non-linear fit of $\sigma(E)$ to a Lorentzian line shape — is required, in addition to the extensive dynamical calculations. The procedure is flawless for isolated resonances, especially if the harmonic inversion algorithms are employed, but the uncertainty of the fit grows as the resonances broaden, start to overlap and 'melt' into the unresolved spectral background. The unimolecular dissociations of most molecules with a deep potential well feature overlapping resonances [133]. It is desirable, therefore, to have robust computational approaches which yield resonance parameters and wave functions without an intermediate fitting procedure, irrespective of whether the resonances are narrow or broad, overlapped or isolated.

4.1.2. Direct approaches

Direct methods, which treat the decay as an eigenvalue problem, often satisfy the requirements mentioned above. The original Hamiltonian, however, cannot be simply diagonalized above the dissociation threshold. As discussed in 2.1, the resonance eigenstates have to fulfill special boundary conditions leaving only outgoing waves at large inter-fragment distances. As a result, the resonance eigenfunctions exponentially diverge as $R \to \infty$ and this makes resonances rather inconvenient objects for a numerical study. For a single decay channel, straightforward incorporation of these diverging waves into a computational scheme is possible only in the close vicinity of the dissociation threshold [113, 187]. At higher energies, the boundary conditions have to be satisfied in a different way.

Physical intuition suggests that positions and widths are mainly determined in the interaction zone of the potential, while the divergent tails of the eigenstates appear in the asymptotic part, where the fragments cease to interact. These considerations are basic to one of the earliest numerical algorithms of a direct estimation of resonance positions, namely the stabilization method [188, 189]. The decaying system is placed in a box and diagonalized in an L^2 basis set of localized functions which ignore the asymptotic behavior of resonances. The diagonalization is repeated many times with varying number of basis functions or box size, and the resonance energies are singled out as those which remain unchanged upon these variations. Although the method provided a clear physical picture of the decay and was later extended to the calculation of resonance widths [190], its practical applications were limited, mostly to one-dimensional systems (see, however, Ref. 191, in which the spectrum of HCO was studied using stabilization).

In the complex-scaling method (reviewed in Refs. 192 and 193) the resonance states are handled with more care. The exponential divergence is suppressed by rotation of the dissociation coordinate in the complex plane, which makes the kinetic energy and the potential complex, but forces eigenstates to become square integrable. A rigorous mathematical justification of this procedure was proposed [194–196], but the method was supposedly used by Hartree and coworkers long before, during the Second World War [197]. For a special class of potentials, complex rotation can be shown to leave the poles of the **S**-matrix invariant, so that the eigenstates of the rotated operator coincide with the Siegert states [192]. For an arbitrary multi-dimensional PES the corresponding theorems are not proved and, moreover, analytic continuation of such potentials to complex coordinate values is far from being straightforward. A version of the method, known as exterior complex scaling, in which the dissociation coordinate is rotated only in the asymptotic region, can be used in such cases [193].

Another way to impose the outgoing boundary conditions on the wave functions is to augment the original Hamiltonian in Eq. (11) by a *complex absorbing potential* [59,164,177,198–203]. This method, already mentioned in connection with the wave packet propagation, is conceptually simple and easy to use, although not as rigorous as the original version of the complex scaling. The driving idea is somewhat similar to the stabilization approach. Since the nontrivial part of the decay occurs only in the interaction zone of the potential, one can artificially damp the wave functions at large interfragment distances in every reaction channel, thus making the eigenstates square integrable and numerically tractable. Simultaneous damping of all states is equivalent to an extra potential term with a negative imaginary part, which differs from zero only at large distances R,

$$\hat{H}_{\text{eff}} = \hat{H} - \mathrm{i}\lambda W(R)\,, \tag{16}$$

where both $\operatorname{Re} W$ and the parameter λ, which controls the absorption strength, are positive. If the radial potential $\operatorname{Im} W \to \infty$ as $R \to \infty$, the spectrum of the non-Hermitian complex symmetric augmented Hamiltonian is discrete and its eigenvalues are complex. The expression above resembles effective Hamiltonians with optical potentials introduced in the context of the Feshbach projection operator formalism [204–206]. It is also similar in spirit to the exterior complex scaling method [193]. Moreover, a similar exponential damping of eigenstates is used in solving the time-dependent Schrödinger equation. Remember that during the time propagation some part of the wave packet is reflected back from this 'complex wall'. In the framework of the time-independent approach this means that, although the eigenvectors of the effective Hamiltonian do approximate the Siegert states of the original Hamiltonian, the two sets of states, the exact one and the approximate one, are different. They coincide in the limit $\lambda \to 0$, as can be rigorously proved for one-dimensional potentials [203,207]. In practice, the limit $\lambda \to 0$ is unattainable and one is forced to resort to numerical experimenting. Different absorbing potentials have been discussed [177, 202, 207], which minimize spurious reflections. The absorption parameters are considered optimal, if the eigenstates remain stable with respect to small variations of $W(R)$ and λ [203, 208].

4.1.3. Basis sets

The essential step of any direct method is the diagonalization of the complex-scaled or effective Hamiltonian in the basis of L^2 basis functions. Rather simple in one or two dimensions, diagonalization can become a challenging problem for three fully coupled degrees of freedom. Chemically bound molecules often have many bound and resonance states (for example, for NO_2 the number of only the bound states is close to 4000), so that extremely large basis sets are required to correctly represent the Hamiltonian matrix. The problem can be made tractable using two different approaches [209]. One can try to optimize the basis set, thereby reducing the number of basis functions. Efficient optimization methods, developed in the context of bound-state quantum calculations of polyatomics [210,211], can be implemented in resonance calculations, too. Often, the method of choice is a contraction-truncation scheme [48, 210], in which the final optimized basis is created after a set of successive pre-diagonalizations of Hamiltonians with lower dimensionality. Alternatively, one can use iterative methods, based on a repeated action of the Hamiltonian on a state vector [181, 185, 208, 212–219]. One of them, the wave packet propagation, has already been described above. However, instead of calculating the autocorrelation function, one generates the matrix elements of the Hamiltonian or the basis functions. As with all iterative methods, explicit storage of the full matrix in the main computer memory is avoided, in contrast to direct diagonalization schemes. Different iterative schemes have been put forward. For example, the Lanczos recursive algorithm [220], which produces a sparse tridiagonal easy-to-diagonalize Hamiltonian matrix, can be used for resonance calculations [181, 216–219]. Iterative procedures built on the Chebyshev expansion of the Green's function [215, 221] also have been developed. Finally, an optimal basis set can be iteratively constructed using filter diagonalization [181, 208, 216–218]. This method, in which few basis functions are adapted to a given small energy interval, is treated in more detail in 4.3.

For most of the direct and indirect methods described above one needs some sort of primary basis, in which the kinetic and potential energies, as well as the state vectors, can be represented. The DVR of Light, Hamilton and Lill [222] proves to be computationally convenient and is widely used in multi-dimensional quantum computations (see Ref. 209 for a recent review). The DVR basis can be visualized as a set of functions each strongly localized around one point in the coordinate space, so that the whole operator can be represented on a discrete space grid. The representation can be obtained, for example, by diagonalizing the position operator in a finite basis set. The DVR is especially useful for orthogonal coordinates, for which the kinetic energy does not contain mixed derivatives (such a choice is always possible for triatomic molecules). In this case, each coordinate can be separately transformed to the DVR. The potential matrix in this representation is diagonal and consists simply of the values of the potential on the grid points. The kinetic energy matrices, defined independently for each degree of freedom, are small and are easily computed in analytic form [223]. The final Hamiltonian matrix can be further used in the contraction/truncation scheme. Moreover, since the resulting Hamiltonian is sparse, the matrix-vector multiplication is fast, which makes the DVR the method of choice for iterative methods.

Calculations of resonance spectra for a rotating triatomic molecule are much more challenging than those for $J = 0$. The full Hamiltonian for $J \neq 0$ [20] is a function of six coordinates: three internal coordinates and three Euler angles, which describe the orientation of the molecule in space. A thorough description of the computational strategies for rotating molecules can be found, for example, in Ref. 20. Calculations are usually done in the non-inertial body-fixed frame of reference which follows the instantaneous orientation of the molecule. The state of the molecule is then characterized by the angular momentum J, its projection M onto the laboratory frame quantization axis, and by the projection K on the body-fixed Z axis (in scattering, the latter is usually taken to be parallel to the inter-fragment vector R). In practice, one uses a mixed representation for the Hamilton operator: The overall angular motion is represented in the functional basis of the rotational Wigner matrices [18], while the internal coordinates are set up in the DVR. Extensions of the DVR grids to the case $J \neq 0$ are discussed by Leforestier [224]. In the mixed representation, the Hamilton super-matrix has a block structure. Each block describes a three-dimensional Hamiltonian with a certain value of K. Since K is allowed to vary between $-J$ and J, there are $2J+1$ such blocks for a given angular momentum J. Note that the 3D Hamiltonians in each block depend on K^2 rather than K, so that the states with $+K$ and $-K$ are degenerate. The adjacent blocks K and $K \pm 1$ are coupled through the coordinate-dependent Coriolis coupling, matrix elements of which form off-diagonal blocks of the super-matrix. Coriolis coupling, arising in the non-inertial body-fixed frame, lifts the degeneracy of the states with $\pm K$. In order to find the spectrum for a given J, one has to simultaneously solve the $2J+1$ coupled three-dimensional problems, and this is exactly what makes the calculations so expensive compared to the case $J = 0$.

The problem is somewhat simplified if one neglects the relatively weak Coriolis coupling and treats each K block independently. This is the essence of the so-called centrifugal sudden approximation [225], which is often the only way to make the calculation feasible, especially for $J \gg 1$. Another approximate scheme for rotating molecules, which is often superior in accuracy to the centrifugal sudden approximation, is the adiabatic rotation method put forward by Bowman [40, 226]. As the name suggests, the method is based on the adiabatic separation of the overall rotation from the internal vibrational motion. The total Hamiltonian is approximated by a sum of the Hamiltonian matrix for $J = 0$ and an additional term $E_K^J(R, r, \gamma)$. The latter is the rotational energy eigenvalue of the symmetric top molecule, i.e., $BJ(J+1) + (A-B)K^2$. This effective potential depends not only on J and K, but also, through the rotational constants, on all three internal coordinates. This method was applied to the calculation of resonance states in HCO [40] and HOCl [227] and very good agreement was found with the exact calculations at low J values. Note that if the rotational constants are frozen in a certain configuration [i.e., $E_K^J(R, r, \gamma)$ is set to a constant] one recovers the well known J-shifting approximation [40]. Examples of calculations of resonance widths for $J \neq 0$ are discussed in Sects. 5 and 6.

4.2. Indirect approach: Kohn variational principle

The principal step in indirect methods is the construction of exact continuum wave functions $\Psi^{(\alpha)}$ in every open fragment channel α for many energies E. An artificial absorption spectrum can then be calculated using Eq. (13) and fitted to a Lorentzian function in the neighborhood of each resonance. Below we describe a workable method for obtaining the wave functions using the variational principle due to Kohn [161]. In the context of reactive atom–diatom scattering, two implementations were developed, one based on the **S**-matrix formalism [223, 228, 229], and another one [158, 159] related to the **R**-matrix theory of Wigner and Eisenbud [162]. We focus on the latter approach, also referred to as the log-derivative version of the Kohn variational principle, and follow the presentation given by Dobbyn et al. [60] and Keller et al. [51], who studied the unimolecular decays of HO_2 and HCO, respectively.

The essential steps of the approach are: (i) splitting the coordinate space into two regions, (ii) solving the Schrödinger equation separately in each of them and (iii) matching the solutions at the dividing surface ∂S. The dividing surface can be taken to be a sphere of radius $R = s$ with its center at the origin (R is the dissociation coordinate); the matching radius s then becomes the parameter of the calculation [162]. If a molecule can decompose in several chemically distinct fragment channels, more elaborate surfaces are necessary [158]. In the outer region, $R > s$, the coupling between the translational motion of the fragments and the internal motion of the diatom BC is weak, so that the interaction between A and BC is a function of the distance R only. In contrast, in the interaction region, $R < s$, the motion is governed by the full non-separable potential $V(R, r, \gamma)$; all energy transfer processes take place there.

The scattering wave function in the outer region in a particular dissociation channel α consists of an incoming wave in this channel, $\imath^{(\alpha)}$, and outgoing waves in all open channels, $o^{(\alpha')}$,

$$\Psi^{(\alpha)} = \imath^{(\alpha)} - \sum_{\alpha'} o^{(\alpha')} S_{\alpha'\alpha} \, , \tag{17}$$

where the amplitudes $S_{\alpha'\alpha}$ of the outgoing waves are the elements of the multi-channel version of the scattering matrix $\mathbf{S}$ introduced in Eq. (5) in 2.1. Asymptotically, the waves $\imath^{(\alpha)}$ and $o^{(\alpha)}$ are products of a plane wave and a wave function ϕ_α of the internal state α of the fragment BC. For short-range potentials and normalized flux they are given by

$$\imath^{(\alpha)}(R, r, \gamma) = o^{(\alpha)\star} \longrightarrow \left(\frac{\mu_R}{\hbar k_\alpha}\right)^{1/2} e^{-\mathrm{i}k_\alpha R} \phi_\alpha(r, \gamma) \qquad \text{for} \qquad R \to \infty \, . \tag{18}$$

Here, the star denotes complex conjugation, and $\hbar k_\alpha = [2\mu_R(E - \epsilon_\alpha)]^{1/2}$, where ϵ_α is the internal eigenenergy corresponding to the eigenfunction ϕ_α.

The waves $\imath^{(\alpha)}$ and $o^{(\alpha)}$ can be calculated on the dividing surface without difficulty for any collision energy E, because the inter-fragment potential for $R \geq s$ is purely radial. The internal eigenfunctions of the BC fragment, $\{\phi_\alpha(r, \gamma)\}$, constitute an orthonormal basis on ∂S [60, 158]. In this representation, the matrices $\langle \phi_{\alpha'} | \imath^{(\alpha)} \rangle$ and $\langle \phi_{\alpha'} | o^{(\alpha)} \rangle$ of the incoming ($\mathbf{I}$) and outgoing ($\mathbf{O}$) waves are diagonal. The matrix $\mathbf{F}$ of the surface values of the partial wave functions $\Psi^{(\alpha)}(R = s, r, \gamma)$ [Eq. (17)] is given by

$$F_{\alpha\alpha'} = I_{\alpha\alpha'} - \sum_{\beta} O_{\alpha\beta} S_{\beta\alpha'} \tag{19}$$

and is non-diagonal. The **S**-matrix is unknown at this stage; it is determined by the dynamics in the interaction region.

Before we discuss the solution in the inner part of the potential, let us first consider the matching conditions at the dividing surface. Suppose a solution $\Psi_<(R, r, \gamma)$ for $R \leq s$ is found. The global wave function in the whole configuration space is obtained by matching the scattering wave in the outer region with the wave function $\Psi_<(R, r, \gamma)$ at ∂S. As in the one-dimensional problems considered in 2.1, we choose to match the logarithmic derivatives at $R = s$. The derivative is taken along the outward normal to the surface, which coincides with R, because the boundary ∂S is a sphere. For each dissociation channel and each point on ∂S one has

$$\left.\frac{1}{\Psi_<}\frac{\partial \Psi_<}{\partial R}\right|_{\partial S} = \left.\frac{1}{\Psi^{(\alpha)}}\frac{\partial \Psi^{(\alpha)}}{\partial R}\right|_{\partial S} = \text{const.} \tag{20}$$

This equation not only establishes the continuity of the log-derivative, but also states that the values of the wave function on ∂S are directly proportional to the values of the normal derivative. Correspondingly, a linear operator $\hat{Y}$ can be formally introduced by

$$\hat{Y}\Psi|_{\partial S} = \partial/\partial R\ \Psi|_{\partial S}\,, \tag{21}$$

which transforms between the wave function and its normal derivative on the surface ∂S. We can express this important relationship using the basis set $\{\phi_\alpha(r, \gamma)\}$ on the surface. For example, for the matrix **F** from Eq. (19) one obtains

$$F'_{\alpha\alpha'} = \sum_{\beta} Y_{\alpha\beta} F_{\beta\alpha'}\,, \tag{22}$$

where the prime indicates the derivative with respect to R. Note, that if *values* of $\Psi_<$ and $\Psi^{(\alpha)}$ are matched on ∂S, their derivatives, calculated using $Y_{\alpha\beta}$, automatically coincide and vice versa. Equations (20) and (22) reveal the appropriate boundary condition for the Schrödinger equation in the interaction region of the potential: One has to look for wave functions with non-zero values on the dividing surface [158].

The matrix **Y**, called the log-derivative matrix, is the central object of the present formulation [60, 158]. For example, the elements of the scattering matrix can be straightforwardly calculated if **Y** is known [158, 230]. This is possible because a unique solution in the inner region — which finally determines the **S**-matrix — is completely defined by the surface values of the wave function.

In the interaction region, $R < s$, one has to find solutions of the Schrödinger equation $(\hat{H} - E)\Psi_< = 0$ with the appropriate boundary condition on ∂S, i.e., $\Psi_< \neq 0$ for $R = s$. The boundary values of $\Psi_<$ are determined from the propagation in the outer region. The wave function in the whole space is then determined using Kohn's variational principle.

Variational principles are widely applied to calculations of bound states in quantum mechanics [18, 231]. One usually considers the expectation value of the energy of the system, $E[\tilde{\Psi}] = \langle\tilde{\Psi}|\hat{H}|\tilde{\Psi}\rangle$, and looks for stationary points (i.e., $\delta E = 0$) of this functional with respect to arbitrary variations $\delta\tilde{\Psi}$ of the trial wave functions $\tilde{\Psi}$. According to the variational theorem [231], the wave function for which the energy functional is stationary is a solution of the Schrödinger equation, while the stationary value itself is the eigenenergy to this solution. The advantage of the variational equation, $\delta E[\tilde{\Psi}] = 0$, is that it can be reduced to a set of algebraic equations, which are easier to solve than the original eigenvalue problem.

Variational formulations can also be developed for the Schrödinger equation in the continuum. However, certain modifications in the construction of functionals are necessary in order to ensure convergence of the integrals. Discussions of the historical development of variational calculations and applications to single- and multi-channel scattering can be found in Refs. 232 and 233.

The functional, relevant for the boundary value problem defined above, was first suggested by Kohn [161]. For a triatomic molecule in Jacobi coordinates it can be written as [60, 158]

$$\begin{aligned}\mathcal{L}\left[\Psi_<\right] &= \frac{\hbar^2}{2\mu_R}\int \mathrm{d}^3\mathrm{x}\, \frac{\partial\Psi_<^\star}{\partial R}\frac{\partial\Psi_<}{\partial R} \\ &+ \int \mathrm{d}^3\mathrm{x}\; \Psi_<^\star \left[-\frac{\hbar^2}{2\mu_r}\frac{\partial^2}{\partial r^2} + B(R,r)\hat{j}^2 + V(R,r,\gamma) - E\right]\Psi_<\,, \qquad (23)\end{aligned}$$

where $\mathrm{d}^3\mathrm{x} \equiv \mathrm{d}R\,\mathrm{d}r\,\mathrm{d}(\cos\gamma)$ denotes integration over the interaction region $R \leq s$. Note that the functional $\mathcal{L}$ is similar to the expectation value of $\left(\hat{H} - E\right)$. However, the kinetic energy in R [the first term in Eq. (23)] acts on $\Psi^\star_<$ and $\Psi_<$ symmetrically. This modification, at the first glance minor, in fact is important, as it makes the modified Hamiltonian Hermitian [60] and allows one to determine the log-derivative matrix $\mathbf{Y}$.

Direct variation of $\Psi_<$ and $\Psi^\star_<$ shows that $\delta\mathcal{L}[\Psi_<] = 0$ is satisfied for such functions $\Psi_<$ which solve the Schrödinger equation and such variations $\delta\Psi_<$, which maintain the surface values of $\Psi_<$, i.e., $\delta\Psi_<|_{\partial S} = 0$. Then, the stationary value of the functional $\mathcal{L}$ is determined by the surface values of the wave function,

$$\mathcal{L}_{\text{stat}} = \frac{\hbar^2}{2\mu_R}\int \mathrm{d}^2\mathrm{x}\ \Psi^\star_< \frac{\partial\Psi_<}{\partial R} = \frac{\hbar^2}{2\mu_R}\int \mathrm{d}^2\mathrm{x}\ \Psi^\star_< \hat{Y}\Psi_< \,. \tag{24}$$

Here, $\mathrm{d}^2\mathrm{x} \equiv \mathrm{d}r\,\mathrm{d}(\cos\gamma)$ is the surface element on the boundary ∂S, the integral is taken over ∂S, and we used the definition of the log-derivative operator $\hat{Y}$ given above. Thus, in just the same way as the standard functionals for bound systems give the estimation of the bound-state eigenenergy, the stationary value of Kohn's functional, Eq. (23), provides the estimation of the log-derivative matrix.

In order to implement the variational method, one needs to introduce an orthonormal basis set to represent the trial wave functions $\Psi_<$ in the inner part of the potential,

$$\Psi_<(R, r, \gamma) = \sum_k c_k v_k(R, r, \gamma)\,. \tag{25}$$

The functional $\mathcal{L}[\Psi_<]$ in this basis takes on the form

$$\mathcal{L}\,[\mathbf{c}] = \mathbf{c}^\dagger\left(\overline{\mathbf{H}} - E\right)\mathbf{c}\,, \tag{26}$$

where $\mathbf{c}$ is the vector of the coefficients $\{c_k\}$, $\mathbf{c}^\dagger$ is its Hermitian conjugate and $\overline{\mathbf{H}}$ is the Hamilton matrix in which the kinetic energy in R is symmetrized as in Eq. (23). The basis set $\{v_k\}$ is defined on the boundary ∂S as well. Remember that for the propagation in the outer region we already introduced another basis set, consisting of the product states $\{\phi_\alpha\}$. Let the set $\{f_\alpha\}$ denote the expansion coefficients of the boundary value of the wave function $\Psi_<$ in the basis $\{\phi_\alpha\}$. The relation between $\{f_\alpha\}$ and $\{c_k\}$ on ∂S is then given by

$$f_\alpha = \sum_k T_{\alpha k} c_k\,, \tag{27}$$

where

$$T_{\alpha k} = \int d^2x \; \phi_\alpha^\star(r,\gamma)\, v_k(R=s,r,\gamma) \tag{28}$$

is a rectangular transformation matrix between the two basis sets on the dividing surface. Note, that the global solution is continuous at ∂S if the f_α coincide with the surface values of the solution in the outer region, $F_{\alpha\beta}$, in each channel β. The derivatives with respect to R, f'_α and $F'_{\alpha\beta}$, also must be equal. This is the essence of the matching condition discussed above. Using Eq. (24) and the basis representations in Eqs. (26) and (27) it is easy to prove the following relationship between the stationary value of the functional $\mathcal{L}$ and $\mathbf{Y}$,

$$\mathcal{L}_{\rm stat} = \frac{\hbar^2}{2\mu_R} \sum_{\alpha\alpha'} f_\alpha Y_{\alpha\alpha'} f_{\alpha'} \,. \tag{29}$$

Stationary points of the functional $\mathcal{L}[\mathbf{c}]$ should be calculated through variation of the coefficients c_k. Kohn's variational principle requires the wave function on ∂S to remain fixed during the variation, $\delta f_\alpha = 0$. In view of Eq. (27), this means that variation of the c_k is subject to the additional condition $\sum_k T_{\alpha k}\, \delta c_k = 0$. The standard way to solve a variational problem with constraints is to use undetermined Lagrange multipliers [234]. A technical realization of this method, which we do not describe here, is given in Ref. 60. Using it, one obtains a compact expression for a set of coefficients $\{c_k\}$ which render $\mathcal{L}[\mathbf{c}]$ stationary, namely

$$\mathbf{c} = \frac{\hbar^2}{2\mu_R} \left(\overline{\mathbf{H}} - E\right)^{-1} \mathbf{T}^\dagger\, \mathbf{Y}\mathbf{f} = \frac{\hbar^2}{2\mu_R} \left(\overline{\mathbf{H}} - E\right)^{-1} \mathbf{T}^\dagger\, \mathbf{f}' \,, \tag{30}$$

where $\mathbf{f}$ is the vector of coefficients $\{f_\alpha\}$, and the last equality follows from the definition of the matrix $\mathbf{Y}$. Now, the stationary value $\mathcal{L}_{\rm stat}$ of the functional can be evaluated from Eq. (26) and compared with the result given in Eq. (29). This yields the following expression for the log-derivative matrix,

$$\mathbf{Y}^{-1} = \frac{\hbar^2}{2\mu_R} \mathbf{T} \left(\overline{\mathbf{H}} - E\right)^{-1} \mathbf{T}^\dagger \,. \tag{31}$$

Equations (30) and (31) complete the construction of the scattering wave function in the continuum. According to Eq. (30), the expansion coefficients over the basis functions in the interaction region are uniquely determined by the boundary values of the normal derivatives $\mathbf{f}'$. On the other hand, in each dissociation channel β one has $f'_\alpha = F'_{\alpha\beta}$ (this is the matching condition). Thus, the derivatives f' can be evaluated, using Eq. (22), from the solutions $\mathbf{F}$ in the outer region by exploiting the matrix $\mathbf{Y}$ from Eq. (31).

The calculation of $\mathbf{Y}$ and $\mathbf{c}$ is the simplest if the matrix $\overline{\mathbf{H}}$ and, hence, the resolvent $\left(\overline{\mathbf{H}} - E\right)^{-1}$ are diagonal. Thus, the first step in the numerical solution is the diagonalization of the modified Hamiltonian and the transformations of Eqs. (30) and (31) to this representation. Next, a set of energies E is selected, for which the wave function and the absorption cross section will be calculated. For each E, the log-derivative matrix $\mathbf{Y}$ is evaluated from Eq. (31), the propagation of the scattering waves $\mathbf{I}$ and $\mathbf{O}$ is performed in the outer region of the potential and the normal derivatives $\mathbf{F}'$ on the dividing surface are calculated from Eq. (22) using $\mathbf{Y}$. This yields the wave function in the interaction region according to Eqs. (25) and (30), and the wave functions $\Psi^{(\alpha)}(E)$ in all open reaction channels are then used in Eq. (13) to calculate the cross section $\sigma(E)$. The computationally most demanding part of the entire calculation is the diagonalization of the operator $\overline{\mathbf{H}}$. However, this step has to be done only once! The subsequent calculations with the diagonal Hamiltonian and the wave propagations in the outer region have to be repeated for each energy E. They are fast, so that the cross section can be determined for many energies [60]. Crucial for the numerical efficiency of the algorithm is the size of the basis $\{v_k\}$ in the inner region. In Ref. 60 the primary DVR basis was optimized using the contraction/truncation scheme described in 4.1.

An application of the method described above is illustrated in Fig. 13 for the dissociation of DCO. This spectrum is the theoretical counterpart to the experimental overview spectrum in Fig. 2. All the structures above the quantum mechanical threshold are resonances. For most of them the positions and widths have been extracted and compared with the experimental data [15]. A more detailed discussion of HCO and DCO is given in 5.1.

4.3. Direct approach: filter diagonalization with absorbing potential

In the approach described in 4.2 the boundary conditions for $R \to \infty$ are explicitly satisfied. The method can also be used for direct calculating resonance states, if one suppresses all incoming waves in the outer region [cf. Eq. (17)]. A direct iterative procedure for such calculations was proposed by Keller and Schinke and applied to HCO [160]. Not surprisingly,

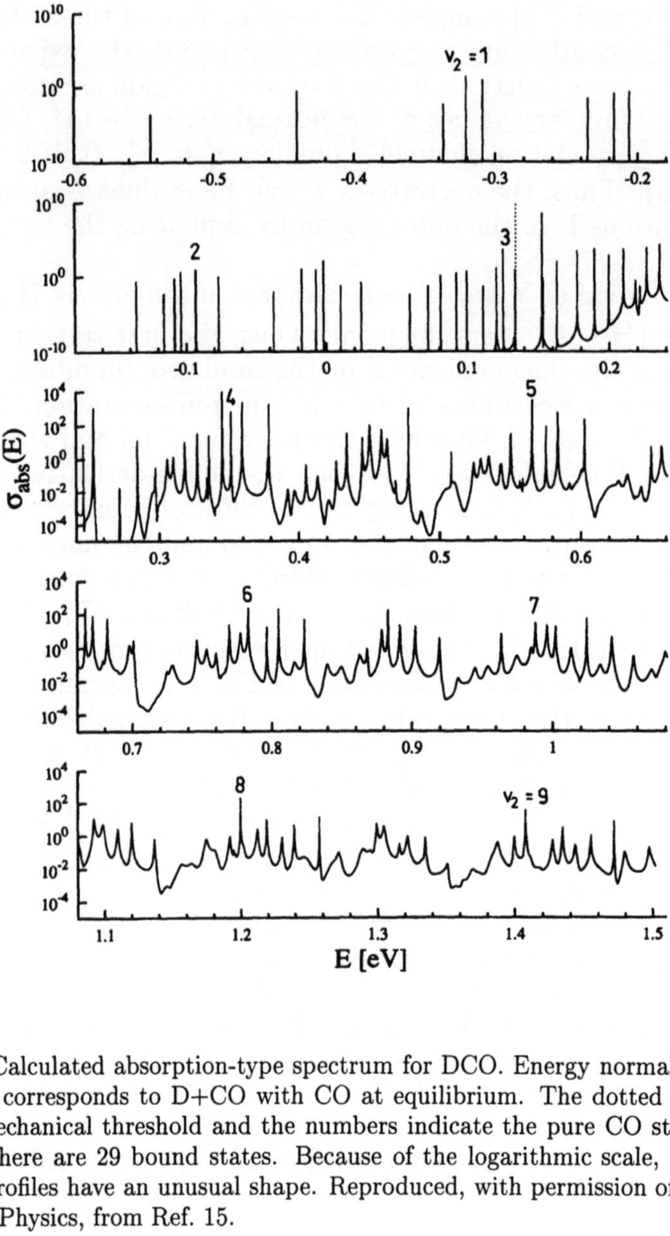

Figure 13: Calculated absorption-type spectrum for DCO. Energy normalization is such that $E = 0$ corresponds to D+CO with CO at equilibrium. The dotted line marks the quantum mechanical threshold and the numbers indicate the pure CO stretching states $(0, v_2, 0)$. There are 29 bound states. Because of the logarithmic scale, the Lorentzian resonance profiles have an unusual shape. Reproduced, with permission of the American Institute of Physics, from Ref. 15.

many more resonances were found than with the method based on analyzing an absorption spectrum. The scheme, which we present in this subsection, is different. It does not impose the boundary conditions rigorously, but uses a phenomenological absorbing potential. Although less exact than the technique described in 4.2, the absorbing potential makes many steps of the variational calculations unnecessary.

In the absorption potential method the metastable resonances are approximated by the eigenstates of the effective Hamilton operator $\hat{H}_{\text{eff}}$, Eq. (16). These eigenstates can be found by diagonalization of the $N \times N$ Hamiltonian matrix in a suitably optimized basis set. However, for many chemically bound molecules the resulting matrices are very large, $N > 10^4$ or so, while the computer time required for the diagonalization grows with N as N^3. The number of basis functions has to be so large, because all eigenstates — starting from the lowest one — are determined in a single calculation, even if one is only interested in some highly excited states. The convergence with respect to the basis size N is enforced across a broad energy window and the only possibility to reach the high energy part of the spectrum is to make this window ever larger, i.e., to use more and more basis functions. As a consequence, this standard approach rapidly becomes impracticable for calculations in the continuum.

The diagonalization could be made much more efficient if one were able to construct a basis set which is *local* in energy, that is, which is appropriate for a narrow energy window $[E_{\text{min}}, E_{\text{max}}]$. Indeed, the required number of basis functions in the window, N_{win}, is of the order of the local density of states, $\rho(E)$, multiplied by $(E_{\text{max}} - E_{\text{min}})$, so that for small energy intervals $N_{\text{win}} \ll N$. An efficient procedure for constructing such a local set of basis functions has been suggested by Neuhauser [181, 216–218] and is called *filter diagonalization* (FD). Suppose that some state χ_0 of the system is chosen, which initially overlaps with many eigenstates over a broad energy range. Suppose further, that a certain operator $\hat{f}(E - \hat{H}_{\text{eff}})$ is introduced, which acts upon χ_0,

$$\Psi_i = \hat{f}\left(E_i - \hat{H}_{\text{eff}}\right)\chi_0 \,, \tag{32}$$

where the real energies $\{E_i\}$ lie between E_{min} and E_{max}. Now imagine that the operator $\hat{f}$ is such that each Ψ_i is a linear combination of only a few eigenfunctions of $\hat{H}_{\text{eff}}$ in the vicinity of E_i. Then, if the grid $\{E_i\}$ is dense enough, the set $\{\Psi_i\}$ constitutes an appropriate basis within the range of the filter and the eigenstates in the window can be obtained by diagonalizing the matrix $\langle\Psi_i|\hat{H}_{\text{eff}}|\Psi_j\rangle$. Note that all basis functions are 'filtered out' of a single initial state χ_0, which explains the name of the method. After the eigenstates in the chosen window are found, one shifts the filter to the next energy window and thus scans the whole energy range in which eigenstates are desired. The essential advantage of the FD method is that, in principal, converged energies and wave functions can be obtained in any region of the spectrum at relatively low cost.

There exist many different operators, which possess the filtering property. For example, Neuhauser [216] originally used a Gaussian function, $\hat{f} \sim \exp[-\beta(E_i - \hat{H})^2]$. When applied to an initial state χ_0, it smoothly suppresses the contributions of eigenstates lying outside the strip $\pm 1/\sqrt{\beta}$ around E_i. Another filtering operator, suggested by Grozdanov, Mandelshtam and Taylor [208], is associated with the damped Green's function [177, 202, 208] built on the operator $\hat{H}_{\rm eff}$, i.e.,

$$\hat{f} = \mathrm{Im}\,\hat{G}^+(E) = \mathrm{Im}\left(E - \hat{H} + \mathrm{i}\lambda W\right)^{-1} . \tag{33}$$

For calculating bound-state energies and wave functions, $\lambda \to 0$ and the filter becomes equivalent to the spectral density operator $\delta(E_i - \hat{H})$ [208]. In the case of resonances, the Green's function in Eq. (33) selects the contributions from those complex poles whose real parts lie near E_i. In what follows we consider only the filter defined in Eq. (33).

The main problem now is to calculate the action of the Green's function onto the initial state χ_0. The standard strategy is to expand $\hat{G}^+$ in a power series of $\hat{H}$. For vanishing λ, a highly efficient expansion exists [215, 235] in terms of Chebyshev polynomials, $T_n(\hat{H})$, which is similar to the one used in short-time wave packet propagations [166, 171]. For finite λ, this expansion has to be modified to account for the absorbing potential. As was shown by Mandelshtam and Taylor [221], the analytically continued Chebyshev polynomials, $T_n^{\hat{\gamma}}(\hat{H})$, can be used for this purpose. If the initial wave packet is localized in the interaction region, outside the range of action of the absorbing potential $W(R)$, the product $\hat{G}^+\chi_0$ takes on the form

$$\left(E - \hat{H} + \mathrm{i}\lambda W\right)^{-1} \chi_0 = (\mathrm{i}\Delta H \sin\phi)^{-1} \sum_{n=0}^{\infty} a_n(E) \left[T_n^{\hat{\gamma}}(\hat{H})\chi_0\right] . \tag{34}$$

Here, both the expansion coefficients a_n and the phase ϕ are energy-dependent, and ΔH is the so-called spectral range of the Hamiltonian, $\Delta H = (H_{\rm max} - H_{\rm min})/2$, where $H_{\rm min}$ and $H_{\rm max}$ are the minimum and maximum eigenvalues of the Hamiltonian in a finite basis representation. The Chebyshev vectors $\xi_n^{\hat{\gamma}} \equiv T_n^{\hat{\gamma}}(\hat{H})\chi_0$ can be iteratively generated from the recursion relation, designed by Mandelshtam and Taylor [221],

$$\xi_0^{\hat{\gamma}} = \chi_0 , \qquad \xi_1^{\hat{\gamma}} = e^{-\hat{\gamma}}\hat{H}\chi_0 , \qquad \xi_{n+1}^{\hat{\gamma}} = e^{-\hat{\gamma}}\left[2\hat{H}\xi_n^{\hat{\gamma}} - e^{-\hat{\gamma}}\xi_{n-1}^{\hat{\gamma}}\right] . \tag{35}$$

In the limit $\lambda \to 0$ we have $\hat{\gamma} \to 0$ and the ordinary Chebyshev recursion relations are recovered [236]. The damping operator $e^{-\hat{\gamma}}$ in Eq. (35) is equivalent to a certain absorbing potential $W(R)$ [208, 221]; therefore, by specifying the function $\hat{\gamma}(R)$ one also fixes the absorbing term in the Green's function. Often, $\hat{\gamma}$ is chosen to be quadratic in R, i.e.,

$$\hat{\gamma}(R) = \hat{\gamma}_0 \frac{(R - R_{\text{min}})^2}{(R_{\text{max}} - R_{\text{min}})^2} \tag{36}$$

for $R \geq R_{\text{min}}$ and $\hat{\gamma} = 0$ for $R < R_{\text{min}}$. R_{min} and $\hat{\gamma}_0$ are parameters of the calculation. One has to ensure that the resonance positions and widths do not depend on them [237]. Note, that it suffices to specify the damping function $\hat{\gamma}$; the absorbing potential W does not appear in the propagation.

The coefficients in Eq. (34) are given by

$$a_n(E) = (2 - \delta_{n0})\, e^{-in\phi(E)}\,, \tag{37}$$

where the phase is defined as $\phi(E) = \arccos([E - \langle H \rangle]/\Delta H)$ (the median of the spectrum, $\langle H \rangle$, is the average of H_{min} and H_{max}). Note that the energy dependence of the Green's function is accumulated exclusively in the coefficients a_n. Such an expansion, which is global in energy, is very convenient, because one and the same set of vectors $\xi_n^{\hat{\gamma}}$ can be used to construct the Green's function at different energies. Moreover, making the energy-to-time Fourier transform of the Green's function and its expansion, one immediately obtains from Eqs. (34) and (35) a representation of the time-evolution operator, which is global in time, i.e.,

$$e^{-\mathrm{i}\hat{H}\,t/\hbar}\,\chi_0 = \sum_{n=0}^{\infty} b_n(t)\, \xi_n^{\hat{\gamma}}\,, \tag{38}$$

with

$$b_n(t) = (-\mathrm{i})^n (2 - \delta_{n0})\, e^{-\mathrm{i}\langle H \rangle t/\hbar}\, J_n(\Delta H\, t/\hbar)\,, \tag{39}$$

where J_n is the Bessel function of order n. Thus, the Chebyshev vectors $\xi_n^{\hat{\gamma}}$ can be used in time-dependent *and* time-independent calculations. In this formulation, the difference between the two approaches becomes visible only at the final stage, when one decides to sum up either the series (34) or (38). The core of the algorithm, Eq. (35), depends neither on time nor on energy.

With Eqs. (32)–(35), one can construct the basis functions $\Psi(E_i)$, filtered out of the initial state near the energies E_i. Before applying the modified Chebyshev expansion, one has to rescale the original Hamiltonian $\hat{H}$, because Chebyshev polynomials converge only for $|x| \leq 1$ [236]. In fact, $\hat{H}$ in the equations above has to be substituted with the operator $\hat{H}_{\text{norm}} = (\hat{H} - \langle H \rangle)/\Delta H$, whose eigenvalues lie between -1 and 1. Then, a basis function can be written as [208]

$$\Psi(E_i) = \frac{1}{2}\chi_0 + \text{Re}\sum_{n=1}^{\infty} e^{-in\phi(E_i)}\, \xi_n^{\hat{\gamma}} \,. \tag{40}$$

Note, that the basis functions remain real if the initial state χ_0 is real.

In practice, an equidistant energy grid $\{E_i\}$ of the size N_{win} is chosen inside a given energy range $[E_{\min}, E_{\max}]$ and all N_{win} basis functions for different E_i are simultaneously calculated from Eq. (40). The number of significant terms in the expansion is proportional to the spectral range ΔH multiplied by the average density of states inside the window. The obtained basis set $\{\Psi_i\}$ is, of course, not orthogonal. In order to construct an orthonormal basis and, additionally, to get rid of redundant linearly dependent functions, the Schmidt orthogonalization procedure is applied to the set $\{\Psi_i\}$. Next, the matrix elements of the effective Hamiltonian, Eq. (16), are evaluated in this basis and the resulting complex symmetric matrix is diagonalized to give a set of eigenvalues and eigenfunctions inside the window $[E_{\min}, E_{\max}]$. The main numerical effort in the calculation is spent on the recursive calculation of the vectors $\xi_n^{\hat{\gamma}}$. Once per iteration, the Hamiltonian has to be applied to a wave function. Since this operation has to be performed very often — 500 000 times is not unusual, — it is the time determining step of the calculation. Therefore, it is essential to calculate $\hat{H}\xi_n^{\hat{\gamma}}$ very efficiently, for example, by using the DVR discussed in 4.1 [176]. Examples of applications of the FD method and its variants are discussed in 5.2 (HOCl) and 6.2 (NO_2).

5. MODE-SPECIFIC DISSOCIATION

Weak coupling between at least one degree of freedom and the reaction coordinate is an essential prerequisite of mode-specific unimolecular dissociation [83, 84]: Energy can be efficiently stored in this particular mode for a long time leading to a small dissociation rate. Excitation of one of the other modes usually increases the energy flow out of the weakly coupled subsystem and increases the fragmentation rate. Weak coupling is often characteristic of molecules with shallow potential wells, e.g., van der Waals molecules. However, a shallow potential is not a necessary requirement. Weak coupling can also be found in molecules with comparably deep potential wells; an example will be discussed in 5.2.

Two molecules with weak intermode coupling and a large degree of mode specificity are HOCl and HCO in their ground electronic states. Both molecules have been intensively studied experimentally as well as theoretically allowing comparisons between measurements and calculations on a high level of detail [126,238]. These studies have significantly advanced the understanding of unimolecular dissociation of small molecules. In this section we will highlight the most important aspects of these studies, starting with HCO, for which the largest body of experimental results is available.

5.1. Dissociation of HCO

The formyl radical is an important intermediate in combustion processes [239] and atmospheric chemistry [240]. As a consequence of its small dissociation energy ($D_0 \sim 0.63\,\text{eV}$ or $\sim 5\,081\,\text{cm}^{-1}$) there are only 15 bound states [51] (29 for DCO [15]). However, the actual number of accessible states for HCO is much larger, consisting of this unusually small number of bound states, even for a triatom, and the many resonance states extending to energies high above the dissociation threshold of the reaction

$$\text{HCO}(\tilde{X}^2A') \longrightarrow \text{H}(^2S) + \text{CO}(X^1\Sigma^+)\,. \tag{41}$$

The longest progression of resonances is built on the overtones of the CO stretching excitation with up to 12 quanta, which corresponds to $\sim 21\,000\,\text{cm}^{-1}$ of excitation energy with respect to the vibrational ground state (0,0,0). In what follows, v_1, v_2, and v_3 are the HC stretching, the CO stretching and the HCO bending mode, respectively.

Long progressions of resonance states were first observed by LIF spectroscopy using the $\tilde{B}^2A'$ state as the intermediate [16]; dispersing the fluorescence yields an overview spectrum, which looks very similar to that of DCO shown in Fig. 2. The large difference in the CO equilibrium bonds, $2.23\,a_0$ in the ground state compared to $2.76\,a_0$ in the excited state [241], makes it possible to observe emission to highly excited vibrational states in the $\tilde{X}$ manifold with many quanta of excitation in the CO mode. In contrast, the HC bond distances are quite similar, so that only states with few quanta in the HC mode, i.e., the dissociation mode, can be observed.

The resolution of the LIF spectra is not sufficiently high to measure linewidths. Precise resonance widths can be achieved only in SEP spectroscopy experiments as performed by Adams et al. [242], Neyer et al. [243, 244] and Tobiason et al. [16]. The work of Tobiason et al. is the

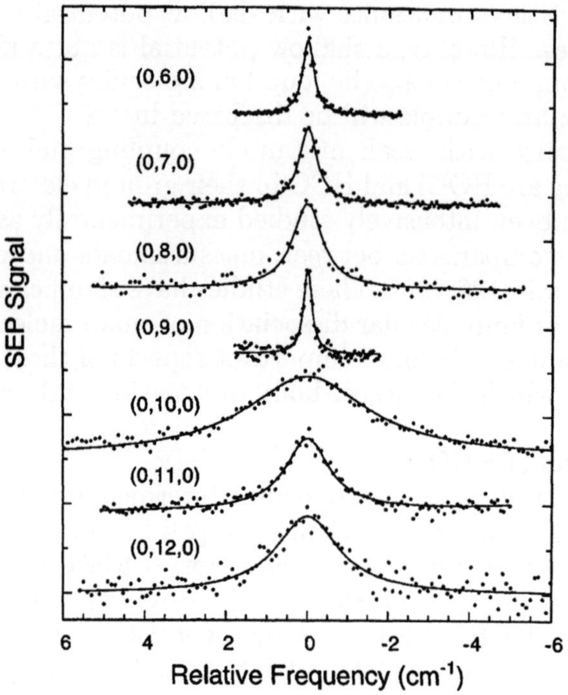

Figure 14: SEP spectra of a series of $(0, v_2, 0)$ resonances of HCO. The solid line in each spectrum is a fit to a single Lorentzian. Reprinted, with permission of the American Institute of Physics, from Ref. 16.

most extensive one with positions for 69 resonance states and linewidths for at least 49 of them. Most of the vibrational levels, especially the narrower ones, can be assigned and therefore it is possible to analyze how the linewidth changes inside a particular progression, e.g., $(0, v_2, 0)$, or between different progressions, e.g., $(0, v_2, 0)$ and $(0, v_2, 1)$. In Fig. 14 we illustrate how Γ varies inside the pure CO stretch progression $(0, v_2, 0)$. First of all, every line can be well fitted by a Lorentzian, which indicates that the linewidth indeed reflects the lifetime rather than incoherent broadening. Second, the linewidth does not show a monotonic increase as function of v_2 or the excess energy, as predicted by any statistical model, but rather changes in a non-monotonic fashion. Not all of the measured resonances are as well separated from other resonances as the states of the pure $(0, v_2, 0)$ progression. Examples of resonances showing different degrees of overlap with neighboring states are given in Ref. 16.

HCO is small enough to allow rigorous quantum dynamical calculations on the ground-state PES without any approximation or model assumption.

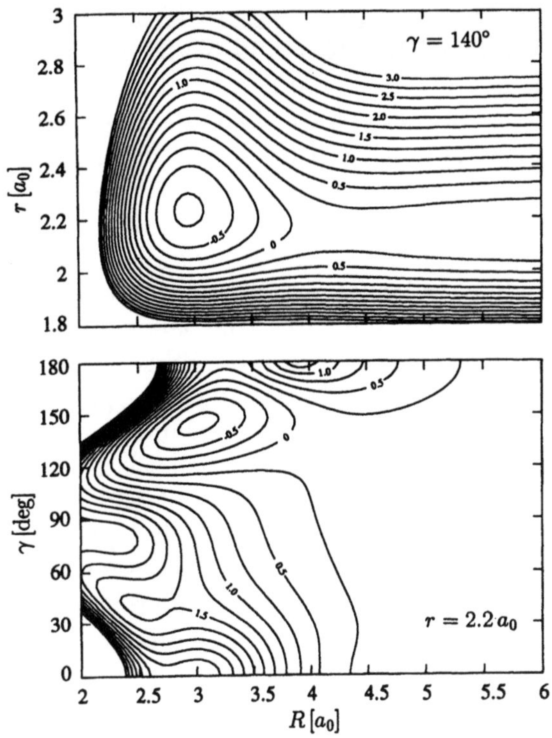

Figure 15: Two-dimensional contour plots of the Werner-Keller-Schinke potential energy surface of HCO. Reprinted, with permission of the American Institute of Physics, from Ref. 51.

Bowman and coworkers were the first to perform such calculations. They calculated a global three-dimensional PES [245] which subsequently was modified in order to better reproduce the available experimental data [246–248]. Much has been learned about the resonance spectrum of HCO from these studies. Nevertheless, with the new results of Tobiason et al. [16] for HCO and later of Stöck et al. [14] for DCO it became apparent that a new PES is needed for explaining the data at high excitation energies.

In the mid nineties Schinke and coworkers constructed a new global PES based on multi-reference configuration interaction (MRCI) electronic structure calculations using a large atomic basis [49]. This ab initio PES was slightly modified in order to correct an underestimation of the CO stretching frequency by $25\,\mathrm{cm}^{-1}$ and an underestimation of the dissociation energy by about $0.05\,\mathrm{eV}$ [51]. The correction of the CO stretching frequency was essential; without this it was impossible to establish a one-to-one correspondence between the measured and the calculated transition energies in the high-energy regime. Figure 15 shows the potential energy contours of the modified PES. R, r and γ are the Jacobi coordinates for H+CO as defined in Sect. 4. There is a small barrier of 0.125 eV in the exit channel, which naturally defines the transition state (TS). Because of the substantial anisotropy of the PES in the vicinity of the TS, the lowest TS frequency is quite large, about $250\,\mathrm{cm}^{-1}$, i.e., the TS is said to be 'tight'. Most important, however, is the observation that the minimum energy paths in the direction of the H+CO channel (R large), on one hand, and the HC+O channel (r large), on the other, are approximately perpendicular to each other. This indicates that the coupling between the HC and the CO stretching modes due to the PES is weak.

With this PES all measured bound-state energies and resonance positions have been reproduced with a root-mean-square error of $17\,\mathrm{cm}^{-1}$; a list of the energies and widths as well as the assignments are published in Ref. 51. The comparison covers a range of 12 CO vibrational quanta. Most of the calculated resonance widths, which depend more sensitively on dynamical details than the energies, also agree very well with the measured ones [51]. In Figure 16 we show for three different progressions the measured and the calculated resonance widths as functions of the CO stretching quantum number v_2; further examples are given in Ref. 51. Such representations are possible because many of the resonances are assignable; if the coupling between the modes were stronger, as for DCO [15], it would not be possible to illustrate how Γ varies with quantum numbers.

The agreement with the experimental widths is generally favourable; most of the trends are satisfactorily described. There are also a few states for which the calculated widths are an order of magnitude off, e.g., $(0,4,1)$ in Fig. 16. We will discuss a possible explanation of this failure below. The widths exhibit a clear-cut mode specificity. The pure CO stretching progression has the smallest widths or the longest lifetimes. Because of the weak coupling between r, on one hand, and R and γ, on the other hand, the energy initially pumped into the CO mode flows very slowly into the dissociation mode. Simultaneous excitation of the other modes drastically increases the widths. Inside the progressions, Γ does not monotonically vary with v_2 but shows pronounced structures. These maxima may extend over only 1–2 quanta or they can be quite broad. The non-monotonic behaviour of Γ as function of v_2 is caused by mixings between different progressions and will be discussed below in 5.3.

In the experimental SEP spectrum only a relatively small number of resonances, generally the narrowest ones, are definitively identified and analyzed. With only very few exceptions, all measured resonances are non-overlapping. The broader resonances with substantial excitation in the dissociation mode overlap; they have small Franck-Condon factors and

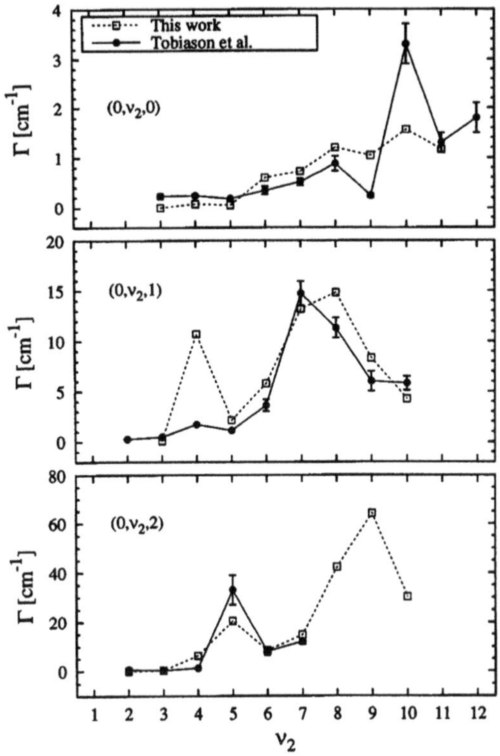

Figure 16: Calculated (dashed lines and open squares) and measured (solid lines and filled circles) HCO resonance widths Γ as function of the CO stretching quantum number v_2 for three progressions. Reprinted, with permission of the American Institute of Physics, from Ref. 51.

therefore low intensities and form the 'background' of the spectrum. The quantum calculation of eigenvalues in the complex energy plane — by using Kohn's variational principle, complex scaling, or absorbing potentials in the exit channels as discussed in Sect. 4 — yields many more resonances than observed experimentally. In the left-hand panel of Fig. 17 we show all the directly calculated widths [160] (ca. 300) as function of the excess energy. Γ fluctuates over more than three orders of magnitude! These fluctuations manifest the mode-specific character of the dissociation of HCO most impressively: Two resonance states may be nearly degenerate; nevertheless, if they have different vibrational assignments, the strengths of coupling to the continuum may be different, too. As was discussed in 2.2, fluctuations in mode-specific decay can be clearly understood within the adiabatic approximation.

At this point it is appropriate to consider the predictions of approximate treatments, that is, classical mechanics and RRKM theory. Proce-

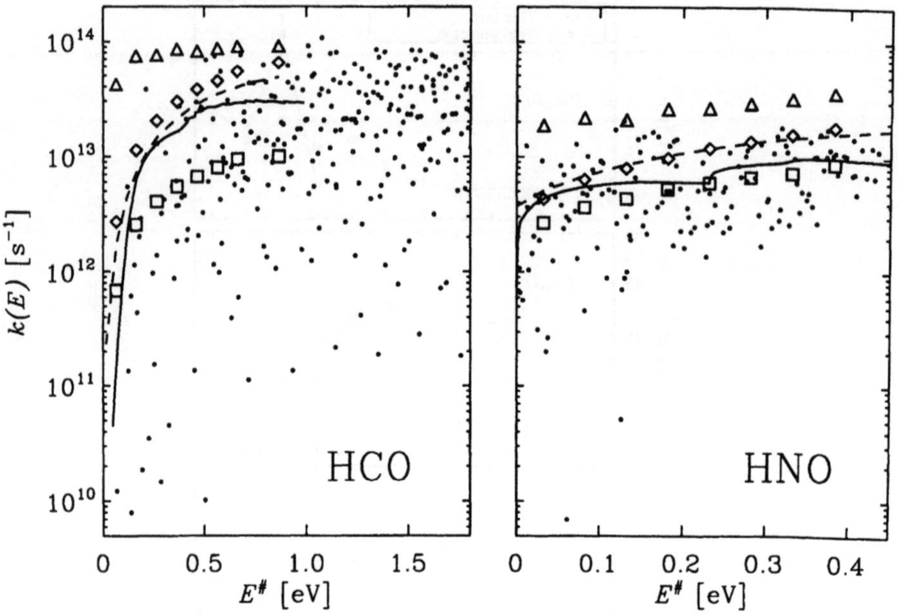

Figure 17: Calculated state-specific rates $k = \Gamma/\hbar$ for HCO and HNO vs. excess energy (small dots). The symbols indicate the fast (triangles), the slow (squares) and the average (diamonds) classical rates (see Sect. 8). The solid and dashed lines are the SACM and the classical RRKM rates, respectively. Reprinted, with permission of the Bunsengesellschaft für Physikalische Chemie, from Ref. 39.

dures for calculating the classical rate will be discussed in 8. The statistical rate is given by the celebrated expression (see Chapters 1 and 2 in this book)

$$k(E) = \frac{N_{\mathrm{TS}}(E)}{h\rho(E)}, \tag{42}$$

where $N_{\mathrm{TS}}(E)$ and $\rho(E)$ are the number of open channels at the transition state and the density of states of the reactant molecule, both evaluated at energy E (see Fig. 1). In the (variational) RRKM theory [249] $N_{\mathrm{TS}}(E)$ is calculated from the classical phase space volume while in the SACM of Quack and Troe [250,251] it is determined using adiabatic potential curves as discussed in 2.2. The RRKM and SACM rates are shown in Fig. 17; the latter is consistently smaller than the former. The order of magnitude of the quantum mechanical rates as well as their average energy dependence are reasonably well described by the statistical rates. The SACM overestimates the arithmetic average of the quantum mechanical rates, but this overassessment is not dramatic. For example, the average quantum rate around an excess energy of 0.5 eV is only a factor of 2–3 smaller than the SACM rate. This is remarkable, since HCO is not a molecule which fulfills the basic assumptions of RRKM theory: The dynamics is mode-specific. The decay of classical trajectories is not exponential, giving rise to at least two classical rates, a fast one and a slow one (see 8.2 for a more detailed discussion); they are both shown in Fig. 17. The average classical rate agrees with the RRKM rate (see Sect. 8).

For comparison, the quantum mechanical, classical and statistical rates for the dissociation of HNO [39] are depicted in the right-hand panel of Fig. 17. The well depth is significantly larger than for HCO and the internal dynamics is much less regular. As a consequence, the breadth of the rate fluctuations is much smaller than for HCO; HNO belongs to the category of statistical state-specific molecules described in Sect. 6. Since for HNO the assumptions of transition state theory are quite well fulfilled — no assignable states near threshold — the SACM rate reproduces the average quantum mechanical rate almost perfectly. The classical RRKM rate is slightly larger than the quantum mechanical SACM rate.

The deuterated formyl radical also has been studied, experimentally [14,252] and theoretically [15,55]. The energies of 98 resonances have been measured and widths have been determined for 90 of them. The agreement between the experimental and calculated energies and widths is similar to HCO [15]. However, there is one important dissimilarity between DCO and HCO: The coupling between the internal vibrational modes is substantially stronger in DCO than in HCO. The frequencies of the two stretching modes differ by only 96 cm^{-1} which gives rise to a systematic anharmonic resonance (not to be confused with scattering resonances) already at low excitation energies [253]. For comparison, the frequency difference for HCO is 572 cm^{-1}. Because of this anharmonic resonance the bound as well as the resonance states are grouped together in quasi-degenerate multiplets, so-called *polyads*. Inside a particular polyad the *local* density of states is considerably higher than it is for HCO, for which there is no anharmonic resonance. It is this increased density of states which causes the stronger mixing in the case of DCO. The consequence is that many states cannot be assigned in an unambiguous manner and therefore relating the magnitude of the decay rate to the quanta in each mode is complicated [15].

5.2. Dissociation of HOCl

Another molecule with mode-specific fragmentation dynamics is hypochlorous acid,

$$\mathrm{HOCl}(\tilde{X}^2A') \longrightarrow \mathrm{Cl}(^2P) + \mathrm{OH}(^2\Pi)\,. \tag{43}$$

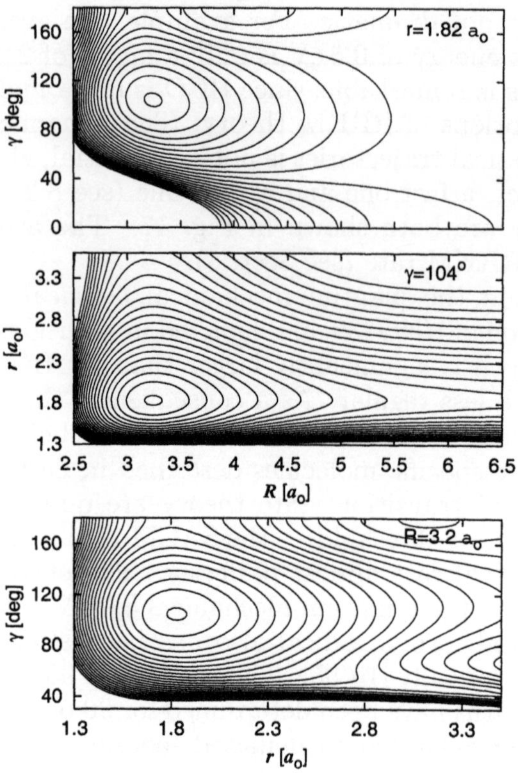

Figure 18: Contour plots of the HOCl ground-state PES. R, r and γ are the Jacobi coordinates appropriate for HO+Cl. Reprinted, with permission of the American Institute of Physics, from Ref. 68.

In this case, it is the HO stretching mode (v_1) which is very weakly coupled to the bending (v_2) and the dissociation mode (v_3). In contrast to HCO, for HOCl the TS is not defined by a barrier: The PES is purely attractive at large Cl+HO distances. The density of states at the threshold is roughly a factor of 20 larger than for HCO [67]. Nevertheless, the mode specificity, reflected by the breadth of the distribution of resonance widths, is even more pronounced.

Two theoretical groups independently tackled the dissociation of HOCl with dynamical methods employing global PES's, which have been calculated at high levels of ab initio theory [67,68,254,255]. Both PES's, which actually are very similar to each other, have been slightly modified in order to reproduce the ca. 20 measured energies of bound vibrational states. Contour plots of the PES are depicted in Fig. 18. There is no potential barrier in the exit channel and the TS is loose, that is, the anisotropy is small in the region of $R \approx 5a_0$, where the O-Cl bond begins to form. Actually, the lowest adiabatic potential curve has no maximum. Remarkable is the extremely weak coupling between the HO stretching and the OCl dissociation modes; the minimum energy paths in the direction of HO–Cl (R large) and H–OCl (r large), respectively, are almost perpendicular to each other. This feature of the HOCl PES and the large OH stretching frequency, compared to the other two frequencies, point to an adiabatic separation for the vibrational degrees of freedom (see 2.2).

Schinke and coworkers used the filter-diagonalization (FD) method to calculate all bound states up to the dissociation limit and resonance states up to an excess energy of $500\,\text{cm}^{-1}$ [68]. An overview of the resonance spectrum up to an excess energy of $4\,000\,\text{cm}^{-1}$ was obtained by harmonic inversion of the autocorrelation function calculated by propagating a random initial wave packet [67]. Bowman and coworkers, on the other hand, employed a truncation/recoupling scheme to calculate bound and resonance states [66, 256]. In all calculations an absorbing imaginary potential has been used in order to restrict the grid in the dissociation coordinate.

In order to understand the dissociation dynamics of a molecule, it is helpful to first analyze how the bound-state spectrum develops with energy from the bottom of the well up to energies close to the threshold [68, 257]. The bending and the OCl stretching frequencies are in a near 2:1 anharmonic resonance, i.e., $\omega_2 \approx 2\omega_3$. The result of this resonance is that the spectrum — for a particular HO stretching quantum number v_1 — is organized in polyads $[v_1, P]$ with polyad number $P = 2v_2 + v_3$. Furthermore, the wave functions exhibit rather complex nodal structures. Contour plots of all wave functions belonging to polyad $[0, 30]$, which is close to threshold, are displayed in Fig. 19. The horseshoe-type shape of the wave function of state (0,0,30) is a direct consequence of the 2:1 anharmonic resonance. The states labeled $(0, v_2, x)_{D(30)}$, which essentially describe evolution of the system along the dissociation route (here, 'D' stands for dissociation), do not exist at low energies, but come into existence at a saddle-node bifurcation at relatively high energies [68, 257, 258]. The 'D' states, whose density quickly becomes larger with energy, are the ones which promote dissociation. Although many of the bound-state wave functions look rather bizarre, for polyad [0,30] all states can be assigned. The assignment becomes obvious when one follows the development from low to high energies. Even though the number of assignable states decreases with energy, there is still a large number of states with clear-cut nodal structure high in the continuum. This is true for the polyads with $v_1 = 0$ and even more so for those with $v_1 > 0$.

As a consequence of the inefficient internal energy redistribution, manifested by the large number of assignable resonance states, the dissociation

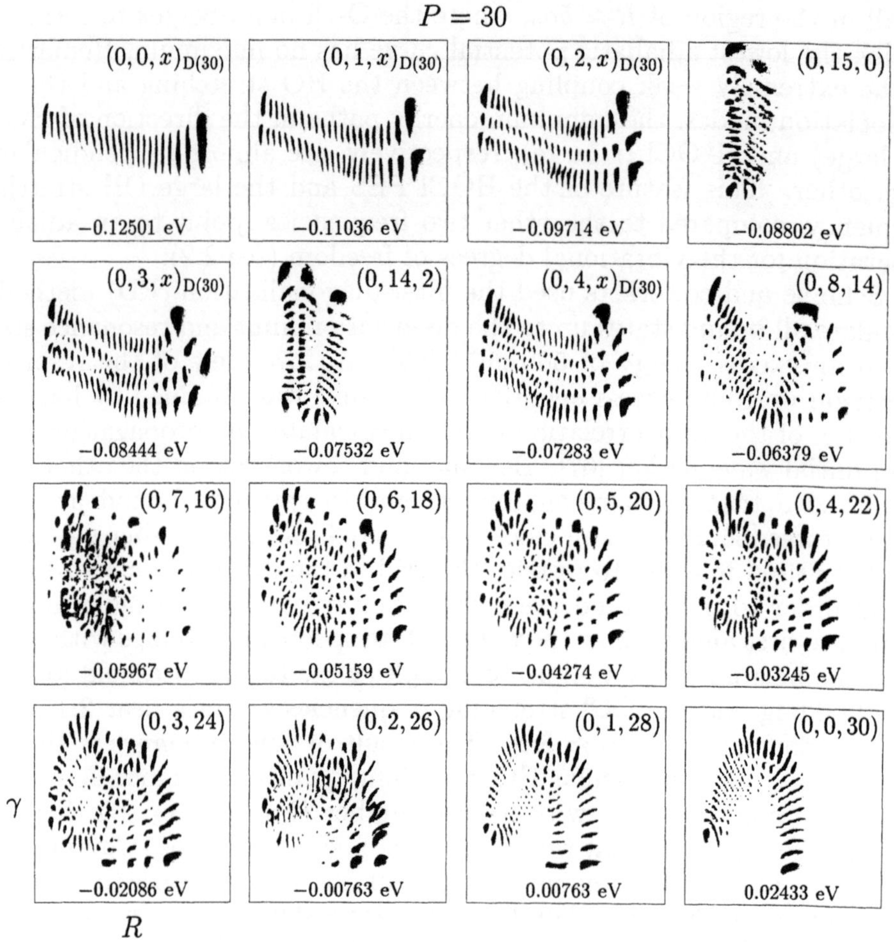

Figure 19: All wave functions for polyad [0,30] of HOCl. Reprinted, with permission of the American Institute of Physics, from Ref. 68.

rates fluctuate over a range of about seven orders of magnitude (Fig. 20). This pronounced degree of mode specificity for HOCl exceeds even that for HCO. The lower bound of the distribution of rates is mainly determined by states with a large number of HO stretching quanta. The qualitative explanation of the mode specificity in terms of an adiabatic approximation has been discussed in 2.2. The SACM overestimates the smallest rates at the lower end of the distribution by many orders of magnitude.

Experimental investigations of the state-specific dissociation of HOCl are limited to the pure overtone 'bright' states (v_1,0,0) which, as discussed in 3.2, can be accessed by direct overtone pumping. Apart from very few

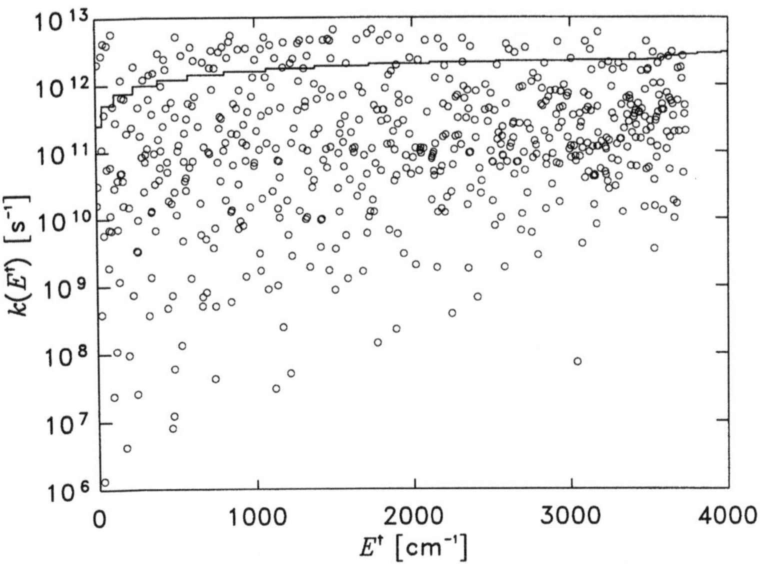

Figure 20: Overview of the calculated dissociation rates of HOCl as function of the excess energy. The solid line is the prediction of the statistical adiabatic channel model (SACM). Adapted from Ref. 67.

exceptions, other states are not accessible by this technique due to unfavorable Franck-Condon factors. A convenient upper state is not present for an SEP experiment. The regions around the bound states with $v_1 \leq 5$ have been spectroscopically analyzed in great detail [259–261]. Actually, the fact that the high overtones of the HO stretching progression can be accessed by overtone pumping already indicates that the HO mode is weakly coupled to the other degrees of freedom.

The highest bound state of the $(v_1, 0, 0)$ progression for non-rotating HOCl is $(6, 0, 0)$; it is bound by about $200\,\mathrm{cm}^{-1}$. Thus, some amount of rotational excitation is required for lifting this state into the continuum. Since the coupling of overall rotation to the internal vibration is usually weak [262], the lifetime is anticipated to be very long. Callegari et al. [27, 263] and Dutton et al. [26] have independently measured the dissociation rate of (6,0,0) for many rotational quantum numbers J and K; here, J is the total angular momentum quantum number and $0 \leq K \leq J$ is the so-called helicity, i.e., the projection of $\mathbf{J}$ on the a-axis, essentially the O–Cl axis (see Sect. 4.1). These measurements provided valuable information about the dependence of unimolecular dissociation rates on the rotation of the parent molecule; similarly detailed information for dissociation on ground-state PES's is not available for other polyatomic molecules.

Examples of measured dissociation rates for $K \leq 5$ and many values

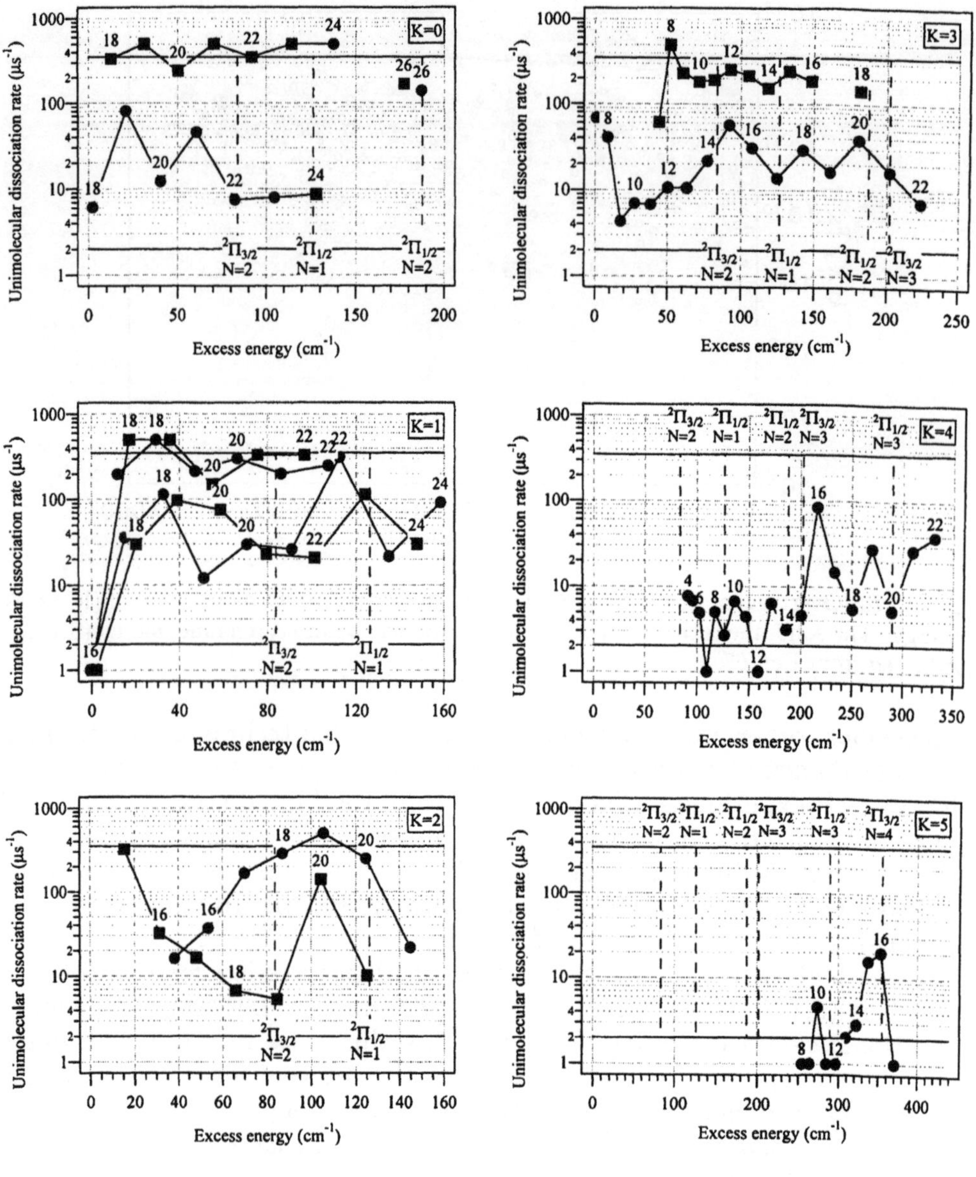

Figure 21: Measured dissociation rates vs. excess energy for vibrational states (6,0,0) and (4,4,2) and different rotational quantum numbers J and K for HOCl. The round/square symbols are used, respectively, for states with mostly (6,0,0)/(4,4,2) character. The horizontal lines show the experimental detection limits. Numbers over the symbols indicate the values of J. Reprinted, with permission of the American Institute of Physics, from Ref. 27.

of J are exhibited in Fig. 21. Several observations can be made. First, the rates are exceedingly small. Second, they do not depend smoothly on J, but show pronounced fluctuations over up to an order of magnitude. Third, the rate averaged over J depends only marginally on the projection quantum number K; both, Callegari et al. [27] and Dutton et al. [26] observed that it slightly decreases with K. Finally, the measured rates are several orders of magnitude smaller than predicted by statistical models (cf. Fig. 20). And this is by no means a surprise: The basic assumption of RRKM theory — randomization of energy before the molecule breaks apart — is not fulfilled for the pure OH overtone states.

In the lower panel of Fig. 22, we compare the experimental rates for $(6,0,0)$ and $K = 0$ with the calculated ones. Because (6,0,0) lies nearly

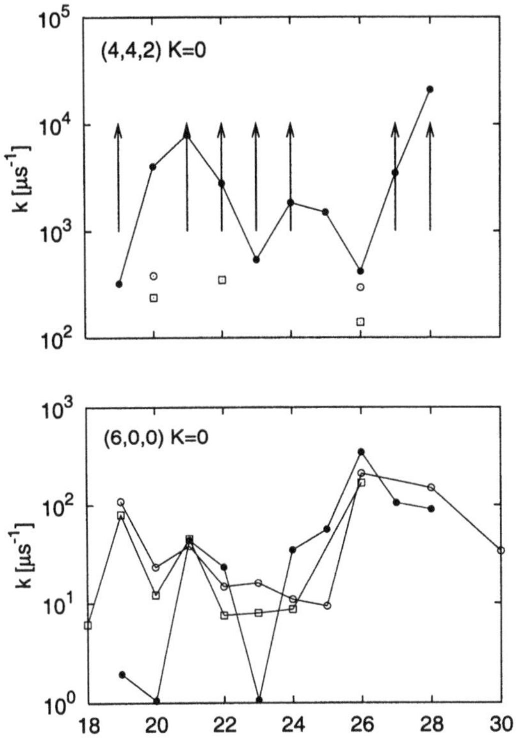

Figure 22: Comparison of measured and calculated (filled circles) dissociation rates for states (4,4,2) (upper panel) and (6,0,0) (lower panel) of HOCl for $K = 0$. The open circles are the results of Dutton et al. [26] and the open squares denote the results of Callegari et al. [27]. The arrows in the upper panel indicate that the widths are larger than $10^3 \mu s^{-1}$ (Ref. 26). Reprinted, with permission of the American Institute of Physics, from Ref. 69.

$200\,\mathrm{cm}^{-1}$ below the dissociation threshold, this state becomes unbound only for $J \geq 18$. The filter-diagonalization calculations [69] have been performed with the best available PES [264], which is a modification (perturbative inversion) of the original ab initio potential; the known experimental bound-state energies are reproduced with this potential within spectroscopic accuracy. Because the rates are so small, they are exceedingly sensitive to subtle inaccuracies of the PES and to all details of the numerical procedure like grid size, absorbing potential, propagation length etc. Resonance positions for different K values lie in different energy windows. Because the computations are very time consuming, they have been performed only for $K = 0$. Coriolis coupling, which is the main obstacle for calculations at high rotational excitation, is approximately taken into account by including only $K = 0, 1$ and 2 states in the expansion of the wave functions.

Despite these numerical difficulties the agreement between experiment and theory is satisfactory. The order of magnitude of the average rate is right and even the broad maximum around $J = 26$ is reproduced by the calculation. The rates for the lowest rotational states, just above threshold, are not trustworthy; here the excess energy is very small and therefore the de Broglie wavelength in the HO-Cl coordinate is very large, which inevitably spoils the absorbing properties of the optical potential. Both, the experimental and the calculated rates show pronounced fluctuations as a function of J. These fluctuations are due to accidental mixings with nearby 'dark' states having $v_1 < 6$. The mixings are tuned by varying J and will be described in more detail in 5.3. The broad maximum around $J = 26$ is due to mixing with state (4,4,2), whose dissociation rate, because of the excitation in the bending and the O-Cl mode, is roughly three orders of magnitude larger than the rate of (6,0,0). The calculated rates for (4,4,2) are compared to the measured ones in the upper part of Fig. 22. Because of insufficient time resolution, rates larger than $10^3\,\mu\mathrm{s}^{-1}$ could not reliably be measured; only lower limits could be given. Considering these experimental limitations, the agreement between calculated and measured rates for (4,4,2) is satisfactory.

Measurements have also been made for vibrational states (7,0,0) and (8,0,0) [265]. The main observation is that on average the rates increase by two orders of magnitude from $v_1 = 6$ to 7 and again by two orders of magnitude from $v_1 = 7$ to 8. Calculations performed in the same way as for (6,0,0) reproduced this strong increase [69]. The calculated resonance widths are depicted in Fig. 23. Because details like the pronounced maxima caused by mixings with other vibrational states are not satisfactorily described by the calculations, only the ranges of the experimental data are indicated. For example, both the theoretical and the experimental widths

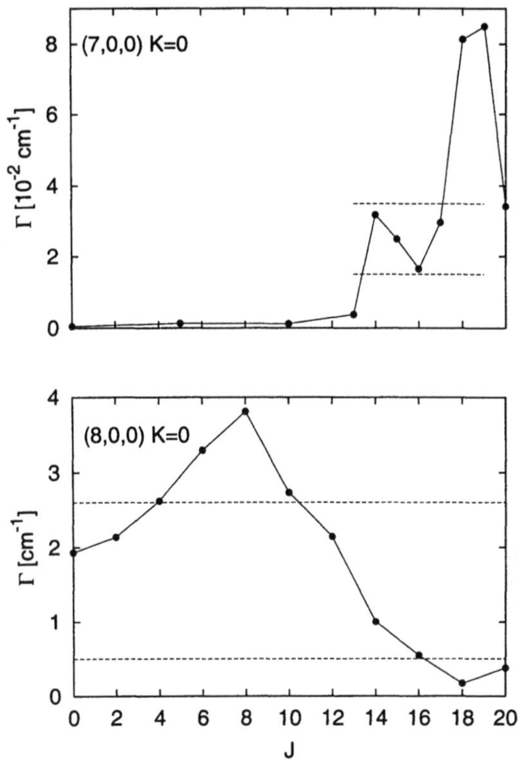

Figure 23: Calculated resonance widths Γ as function of the rotational quantum number J for vibrational states (7,0,0) and (8,0,0) of HOCl. The helicity is $K = 0$. The two dashed horizontal lines indicate the range of the measured widths, Ref. 265. Reprinted, with permission of the American Institute of Physics, from Ref. 69.

for $(8, 0, 0)$ show a resonance maximum. However, while it occurs at $J = 8$ in the calculations, it shows up at $J \sim 15$ in experiment. A possible explanation for the rapid increase of the width from $v_1 = 6$–8 has been given in Ref. 69 in terms of a vibrationally adiabatic representation. Incidentally, we note that the calculated widths for $(9, 0, 0)$ are of the same order as the $(8, 0, 0)$ widths.

5.3. Mixing between resonances tuned by rotational excitation

In the preceding discussion we observed non-monotonic behaviour of Γ as either a vibrational quantum number, e.g., v_2 in HCO (Fig. 16), or a rotational quantum number, e.g., J in HOCl (Figs. 22 and 23), was varied. The origin of these structures is rather universal and was already discussed in 2.3: the mixing between resonances. Two resonances with widths Γ_1 and Γ_2 ($\Gamma_1 \ll \Gamma_2$) may interact, i.e., the narrow state acquires character of the broad one and vice versa. As a consequence, Γ_1 increases while at the same time Γ_2 slightly decreases. According to perturbation theory, the extent of this mixing is inversely related to the energy separation of the two states involved, ΔE. In 2.3 the energy gap between the narrow and the broad resonance was tuned by changing a parameter of the square-well potential and two limiting cases were highlighted.

In a real molecule of course, the PES cannot be changed at will. Nevertheless, mixings are well known in spectroscopy and can be observed in quantities like rotational constants, intensities or any other characteristic feature of an absorption or emission spectrum. The external parameter is usually one of the rotational quantum numbers, J or K. The energy of a rovibrational state is approximately given by an expression for the symmetric top molecule, i.e.,

$$E^{J,K} \approx E^{0,0} + BJ(J+1) + (A-B)K^2\,, \tag{44}$$

where B and A are the two rotational constants. The important point here is that the rotational constants are different for each vibrational state; they depend on the structure of the underlying wave function. As a consequence, the energy separation between two vibrational states,

$$\Delta E \approx \Delta E^{0,0} + \Delta BJ(J+1) + (\Delta A - \Delta B)K^2\,, \tag{45}$$

where $\Delta E^{0,0}$ is the energy difference of the rotationless states and ΔB and ΔA are the differences of the two rotational constants, can be altered by changing J and/or K. For HCO as well as HOCl, A is much larger than B. Thus, K can be used for coarse tuning while fine tuning can be achieved by variation of J.

Examples illustrating how Γ changes as function of the rotational state are depicted in Fig. 24 for HCO. Shown are the widths for six vibrational states belonging to the $(0, v_2, 1)$ progression as functions of K [127]. Because the average energy spacing between neighboring vibrational states is relatively large for HCO, variation of J does not have a big influence on the width. However, changing K leads to pronounced resonance effects. The dependences of Γ on K are qualitatively similar to the dependence of Γ on a in the model potential discussed in 2.3. As mentioned in Ref. 127, inspection of the wave functions does not give a clear-cut indication of which states perturb the $(0, v_2, 1)$ progression. On the basis of the energy spectrum and the expected trends of the rotational constants it was argued that the perturbing states very likely belong to the $(2, v_2-2, 0)$ progression, which, because of the direct excitation of the dissociation mode, have much larger widths, in the range of $100\,\mathrm{cm}^{-1}$. Incidentally we note that the perturbation due to the background states explains the two maxima seen in the middle part of Fig. 16 for $J = 0$ [127].

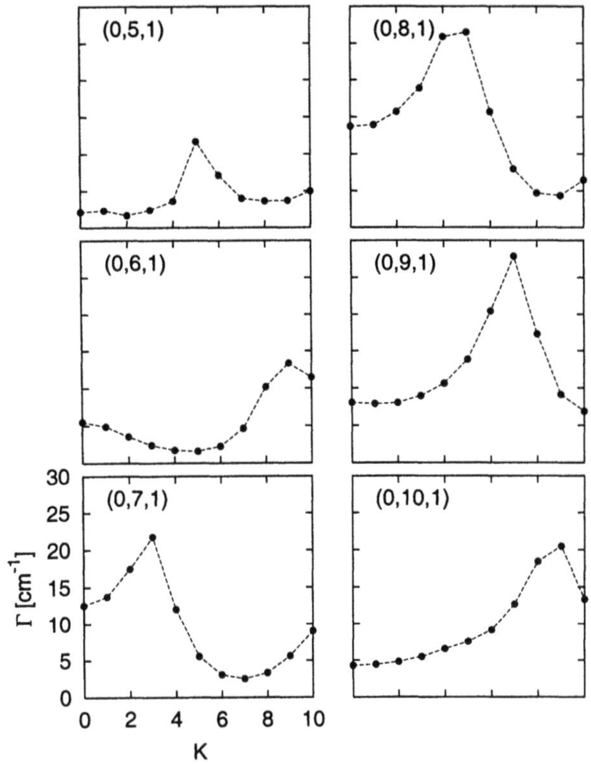

Figure 24: The HCO linewidths for vibrational states $(0, v_2, 1)$ as functions of the rotational quantum number K. In all cases, J is equal to K. The scales are the same in all panels. Reprinted, with permission of the American Institute of Physics, from Ref. 127.

An intriguing example for mixing of resonance states and its influence on the resonance widths is the predissociation of HNO in the first excited singlet state, $\tilde{A}^1A''$, which has been investigated by high-resolution spectroscopy by Pearson et al. [266]. Because the corresponding PES has a high barrier in the H+NO exit channel, the low-lying vibrational states, which are located below this barrier, would have almost infinite lifetimes; tunneling through the high and broad barrier is very unlikely. However, the $\tilde{A}$ state is coupled by Renner-Teller coupling — a particular type of non-Born-Oppenheimer, that is electronically non-adiabatic coupling — to the ground electronic state, $\tilde{X}^1A'$, via which HNO eventually breaks apart [128].

In the absence of any resonance effect, a quasi-bound state of the $\tilde{A}$ manifold is coupled only to the broad and diffuse scattering states, which constitute the continuum of the ground electronic state. Because this con-

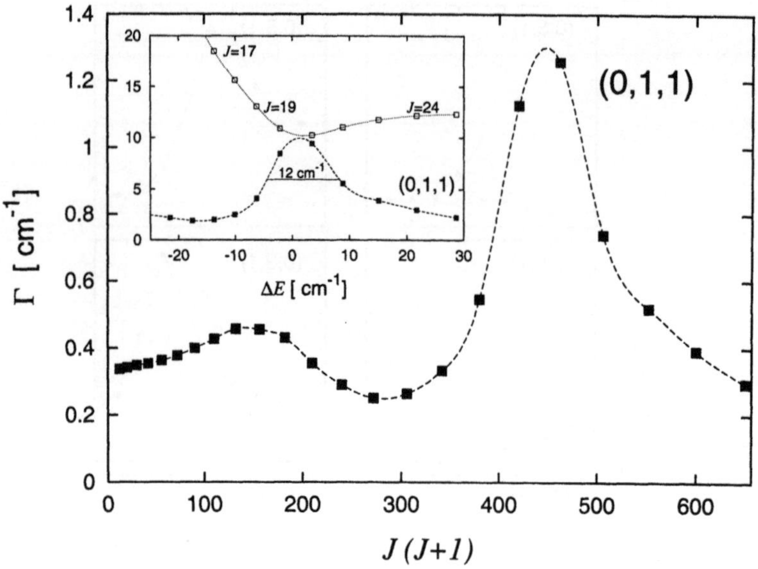

Figure 25: The linewidth of the vibrational state (0,1,1) in the $\tilde{A}^1A''$ electronic state of HNO for $K = 3$. The inset shows the width as function of the energy gap, ΔE, between the bright state (0,1,1) (filled squares) and the perturbing state belonging to the ground electronic state (open squares). In the inset, the (0,1,1) resonance width is multiplied by 7.5. Reprinted, with permission of the American Institute of Physics, from Ref. 128.

tinuum is unstructured, the width does not vary much with J. However, if there is a nearby resonance state in the $\tilde{X}$ continuum, which is localized in the potential well and consequently has a width significantly smaller than the background states, the widths of these two states change as they are tuned into resonance by varying J. An example is shown in Fig. 25. Depicted is Γ for the vibrational state (0,1,1) in the $\tilde{A}$ state with $K = 3$. There is a small maximum around $J = 11$–12 and a more pronounced maximum around 21. The latter is explained by mixing with a much broader state belonging to the $\tilde{X}$-state manifold having a large degree of excitation in the NO stretching mode. The inset shows the widths of both states as function of the energy separation and it is clearly seen that around $\Delta E \approx 0$ the width of state (0,1,1) has a maximum while at the same time the width of the ground-state resonance has a slightly indicated minimum. The behavior in Fig. 25 corresponds to the weak coupling limit illustrated in Fig. 7, i.e., the positions cross, while the widths show an avoided crossing. The coupling between the two manifolds of resonance states is the Renner-Teller coupling, which indeed is weak for HNO, because both electronic states are bent [128]. The breadth of the maximum approximately resembles the resonance width of the perturbing state. It must be underlined that the perturbing state has no clear-cut assignable nodal structure. Nevertheless, it can be recognized as J is varied across the resonance.

This kind of resonance structure in the predissociation of HNO has also been seen in experiment [266], however, at values of J which differ considerably from those obtained in the dynamics calculations. A quantitative description of these mixing effects requires very accurate PES's, because the level structure — in the case of HNO the level structure of two different electronic states — has to be precisely reproduced. The two calculated potentials for HNO do not meet this requirement; for example the vertical energy separation is wrong by about 0.1 eV.

The structures in the rates of HOCl, Figs. 22 and 23, are also caused by resonance effects. If the rotational constants B of the two interacting states differ strongly, only small changes in J are needed to tune in and out of resonance as seen in the low-J region for state (6,0,0). If the nodal structures of the two states are not too different, like (6,0,0) and (4,2,2), the corresponding rotational constants nearly coincide with the result that the resonance extends over several values of J. The maximum for (6,0,0) around $J \approx 26$ and the maximum for state (8,0,0) at $J = 8$ are good examples. Actually, the $J = 26$ maximum for (6,0,0) in Fig. 22 illustrates the strong coupling limit of Fig. 7, that is, the widths of (6,0,0) and (4,4,2) have almost the same value at the resonance maximum indicating that they cross. Although the HOCl PES employed in the most recent calculations [69] is very accurate, the accuracy is not sufficient to guarantee that the resonance structures occur at the correct positions (see the discussion of Fig. 23).

Mixing between resonances belonging to either different electronic states (HNO) or different vibrational manifolds (HCO and HOCl) are general effects; they are the rule rather than the exception. In order to simplify the following discussion, we distinguish three types of resonance states: the narrow states, the broad background states, and the perturbing states. The perturbing states have widths which are, on one hand, substantially larger than the widths of the narrow resonances, but still much smaller than the widths of the strongly overlapping background states. Even though they might not be assignable, in contrast to the background states their wave functions have a clear identity, i.e., they can be recognized and followed when a parameter in the Hamiltonian is varied.

Normally, a narrow resonance is coupled with the dense set of background resonance states. Because the background states look alike, the coupling between the narrow state and the background is not specific and the variation of a parameter, for example J, does not appreciably change the width of the bright state. However, if the narrow state is close to a perturbing state its width can be drastically altered by changing the energy separation. Thus, the general behavior of a resonance width as function of J depends on the mean energy separation between the possible perturber states and therefore on their density. The density of resonance states is very small for HCO and therefore the width varies smoothly with K. The variation of Γ with J is even marginal. On the other hand, the density of possible perturbing states is much larger for HOCl, at least near threshold, and as a consequence Γ shows a more unpredictable variation with J, fluctuations rather than a monotonic dependence (Fig. 22 for $v_1 = 6$). However, at much higher energies, in the region of $v_1 = 8$ there are substantially less resonance states, which might serve as perturbing states, and therefore the (8,0,0) width shows a much smoother behavior (Fig. 23). Of course, there is also the possibility that a narrow resonance is coupled to another narrow resonance, which can lead to interesting effects. From what has been discussed in this section, it is clear that it is impossible to analyze this kind of mixing effects for strongly irregular systems like NO_2, for example.

5.4. Epilogue

At the end of this section it is appropriate to make some general comments concerning HCO and HOCl. In both cases it is sensible to follow how Γ varies from state to state. The densities of states are low enough to guarantee that there are many non-overlapping resonances. The majority of them have distinct nodal structures and therefore can be identified, when, for example, J is changed. Many of the resonances even can be assigned in terms of quantum numbers. In principal, most of the resonances can be resolved experimentally and HCO and DCO are illuminating examples; that only a few resonance states have been investigated for HOCl, has merely a technical reason. This allows one to compare calculated and measured widths on a state-by-state level. The agreement is, except for a few states, excellent for HCO and DCO and somewhat less good for HOCl, although the potential energy calculations have been performed on similar levels of accuracy. The reason is the higher density of states for HOCl. Because the average spacing between neighboring states is smaller for HOCl, mixing effects between resonances are more pronounced. In order to correctly describe these subtle effects, the PES has to have a higher accuracy — an accuracy which cannot be obtained with up to date computer resources. State-by-state comparisons between experiment and theory for other mode-specific molecules, especially molecules not containing hydrogen atoms, would be valuable. Ozone in its ground electronic state is an interesting candidate [42]. For the systems, which we will discuss in 6, comparison with experimental data on a state-by-state level is not feasible.

6. STATISTICAL STATE-SPECIFIC DISSOCIATION

The molecules highlighted in this section, behave dramatically differently than the ones discussed in Sect. 5. Due to the strong coupling between all degrees of freedom, at energies near the dissociation threshold at best very few states may be assignable. The wave functions of almost all resonance states have no clear nodal structure; they cover more or less uniformly the region of the potential well and at first glance look all alike (see the wave function for NO_2 in Fig. 5 for an example). A vibrationally adiabatic decoupling scheme is not helpful. As a consequence, the widths show a pronounced state dependence, but no mode specificity.

In the following, H_2CO and NO_2 are discussed in detail. For both systems there are large sets of experimental data for comparison with either dynamical calculations or statistical models. Both systems have been very helpful in shaping our present-day understanding of unimolecular dissociation in the statistical regime.

6.1. Dissociation of H_2CO and D_2CO

Deuterated formaldehyde has provided unique opportunities for studying unimolecular dissociation dynamics at the level of complete quantum-state resolution [7]. The pioneering Stark level-crossing spectroscopy experiment of Moore and coworkers was the first demonstration of widely fluctuating dissociation rates in tiny energy windows near the dissociation threshold [9,10]. The photophysical process employed to prepare highly excited quantum states of H_2CO in the lowest singlet state, S_0, is illustrated in Fig. 26. Photo-absorption prepares a rotationally resolved low-lying vibrational level in the S_1 electronic state, which is close to the top of the high barrier of the S_0 PES. The S_1 state undergoes radiative decay ($k_{\rm rad}$) as well as non-radiative decay ($k_{\rm nr}$) by internal conversion to highly excited rovibrational S_0 states, which dissociate on the ground-state PES ($k_{\rm uni}$),

$$\begin{aligned} \mathrm{H_2CO}(S_1) &\xrightarrow{k_{\rm rad}} \mathrm{H_2CO}(S_0) + h\nu \\ &\xrightarrow{k_{\rm nr}} \mathrm{H_2CO}(S_0)^* \xrightarrow{k_{\rm uni}} \mathrm{H_2} + \mathrm{CO}\,, \end{aligned} \tag{46}$$

where $H_2CO(S_0)^*$ represents energized H_2CO in the ground electronic state. The lifetime, $\tau = (k_{\rm rad} + k_{\rm nr})^{-1}$, of the excited rovibrational state of the S_1 manifold has been directly measured by observing the time dependence of the fluorescence intensity; for each level it has been fitted by a single exponential [10]. While $k_{\rm rad}$ remains nearly constant for S_1 states of similar

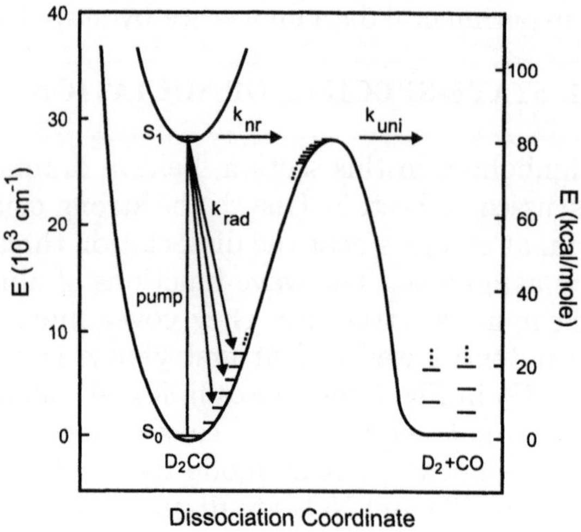

Figure 26: Potential energy diagram for D_2CO. An initially prepared S_1 rovibronic state can undergo both radiative (rad) and non-radiative (nr) decay. Non-radiative decay consists of internal conversion to highly excited S_0 levels followed by unimolecular (uni) dissociation into products D_2 and CO. Redrawn from Ref. 10.

energy, k_{nr} is greatly enhanced when one of the numerous highly excited rovibrational states of S_0 is energetically close — as was described in 5.3 for HNO. Thus, by measuring the lifetime of the S_1 states one can obtain information about the level structure of the S_0 states.

In view of the high density of vibrational S_0 states ($\rho = 437 \pm 112$ per cm^{-1} [10]) variation of the rotational quantum numbers J and K is a rather coarse means for tuning the S_1 states into resonance with S_0 states (as discussed in 5.3). The number of S_0 states, which are indirectly observed through measurement of the lifetime of the few S_1 states, is drastically increased by exploiting Stark level-crossing spectroscopy [10]. Because the dipole moments are different in the two electronic states, varying the strength of an external electric field can be used to continuously alter the energy difference between an S_1 state and the many nearby S_0 states — in contrast to the discrete variation when J is changed. The lifetime is measured for each field strength yielding a Stark level-crossing spectrum for one particular rovibrational S_1 state as illustrated in Fig. 27 for H_2CO and D_2CO. The resonances for H_2CO severely overlap rendering the extraction of individual widths impossible. Because of differences in zero-point energies and tunneling rates, D_2CO decays more slowly than H_2CO and therefore is much better suited for studying the properties of the highly excited S_0 states. Each structure in the figure reflects a rovibrational S_0 state. Because the parameter in the Hamiltonian (the external

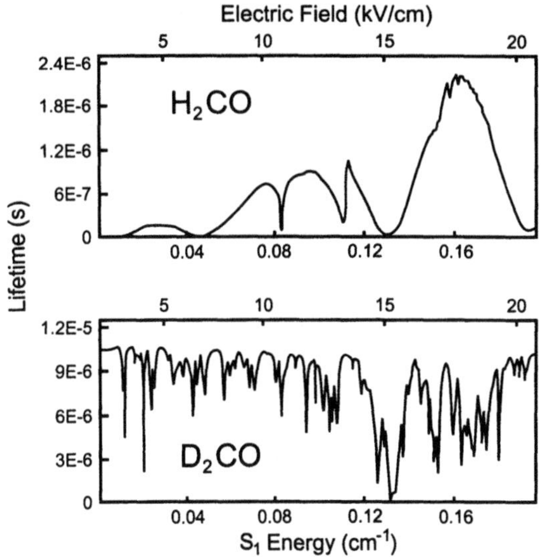

Figure 27: Stark level-crossing spectrum for H_2CO and D_2CO. The vibrational S_1 state is 4^1 (one quantum of mode 4) and the rotational state is $1_{10}(1)$. The energy axis in both plots is the Stark-shift energy of the 4^1 state of S_1. Redrawn from Ref. 10.

field strength) can be tuned continuously, many S_0 states can be observed in a tiny energy interval.

The unimolecular dissociation rates are determined from the fit of the Stark spectra to the expression [267]

$$\Gamma(E_1) = \Gamma_1 + \sum_n \Big[\frac{|\nu_{1n}|^2 (\Gamma_n - \Gamma_1)}{(E_n - E_1)^2 + (\Gamma_n - \Gamma_1)^2/4} \cos 2\theta_n + \frac{2\,|\nu_{1n}|^2 (E_n - E_1)}{(E_n - E_1)^2 + (\Gamma_n - \Gamma_1)^2/4} \sin 2\theta_n \Big], \quad (47)$$

where E_1 and Γ_1 are the energy and the radiative width of the S_1 state, E_n and Γ_n are the energies and widths of the S_0 states, and $|\nu_{1n}|$ and θ_n are the $S_1 - S_0$ coupling matrix element magnitudes and phase angles, respectively; the summation over n extends over all measured S_0 states. Through such fits of the Stark level-crossing spectra one obtains the S_0 resonance widths, Γ_n, as well as the coupling elements $|\nu_{1n}|$ and the phases θ_n. Equation (47) is based on the assumption that the coupling between S_0 and S_1 states is weak, which effectively means $k_{\text{nr}} \ll k_{\text{uni}}$.

Experiments have been performed for the S_1 vibrational states 4^0, 4^1, and 4^2 (zero, one, and two quanta of out-of-plane vibration, mode 4) and

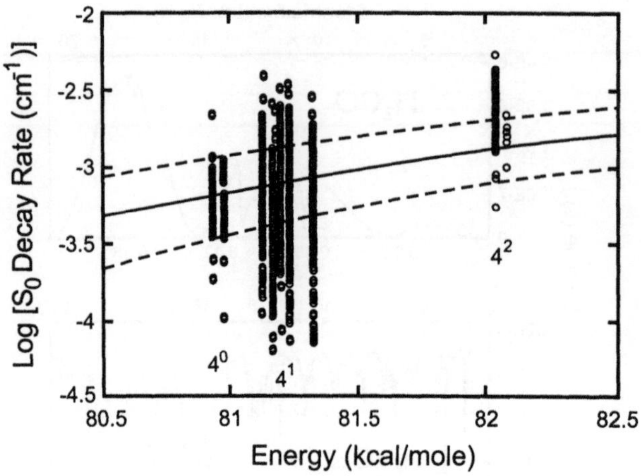

Figure 28: D_2CO S_0 decay widths as a function of energy above the S_0 vibronic origin. Points represent observed widths while the solid line is the width predicted by RRKM theory. The dashed curves represent the uncertainty interval in the statistical calculations. Redrawn from Ref. 10.

rotational states $J = 1$–4; the other modes are not excited in S_1. For 4^1 more than 1200 highly excited S_0 states have been observed; this includes also states with different laboratory-frame projection quantum numbers M_J, whose degeneracy is lifted by the external field. The resonance widths are displayed in Fig. 28 as a function of the energy measured from the vibronic origin of the S_0 state. The authors of Ref. 10 remark: "The two orders of magnitude variation in dissociation rate constants ... observed within the narrow $0.2\,\mathrm{cm}^{-1}$ energy range of a Stark level-crossing spectrum clearly demonstrates that chemical properties of neighboring eigenstates exhibit remarkable state-specific chemical behavior." In other words, the large fluctuations in the dissociation rate reflect wave functions which are considerably different for each of the rovibrational states in S_0. These results may have been surprising at the time they were obtained, but they are now quite understandable in the light of the quantum calculations performed in the last decade. The two orders of magnitude variation of the measured D_2CO rates was found astonishing in 1990; however, this variation is small compared to HOCl or other less statistical systems [78].

No quantum mechanical dynamics calculations of the type described for HCO or HOCl in 5.1 and 5.2 or NO_2 in 6.2 below have been performed for formaldehyde. The reason is not only the lack of an accurate global PES, but also the high density of states near the dissociation threshold. Present-day numerical methods and computer facilities are not yet powerful enough to allow exact calculations for this range of density of states. Therefore comparison with theory is limited to the RRKM prediction for the average rate and the random matrix theory prediction for the distribution of rates to be discussed in 6.3. The RRKM rate (Fig. 28) has been calculated with scaled ab initio transition-state frequencies [10] and it goes right through the experimental data; the uncertainty reflects the uncertainty in the value for the dissociation barrier. The good agreement with the average experimental rate indicates that the assumptions of RRKM theory are fulfilled, i.e., the energy is randomized before the molecule breaks apart. Moreover, the distribution of the experimentally determined coupling matrix elements $|\nu_{1n}|^2$, which essentially reflect the behavior of the S_0 wave functions, is well described by a (Porter-Thomas) χ^2-distribution according to random matrix theory (see 6.3). It is worthwhile noting the statement [10] "... that the observed S_0 unimolecular decay rates fluctuate about the RRKM predicted average value by many times the expected increase with vibrational energy." We will come back to this observation when discussing in 6.4 the possibility of 'steps' in the rate constant.

6.2. Dissociation of NO_2

NO_2 is a benchmark system for experimental as well as theoretical (mainly statistical) investigations of unimolecular dissociation [6]. The spectroscopy, photodissociation, and kinetics have been studied in great detail by many research groups (see Ref. 133 for a review). For recent kinetic studies see Hippler et al. [268] and Harding et al. [269] and references therein. The preparation of highly excited states in the region of the threshold has been described in 3.1 (Fig. 8). It involves photo-excitation of the first excited state, $\tilde{A}^2B_2$, and fast internal conversion to the ground electronic state. The final fragmentation occurs on the ground-state PES:

$$NO_2(\tilde{X}) + h\nu \rightarrow NO_2(\tilde{A}) \rightarrow NO_2(\tilde{X})^* \rightarrow O(^3P) + NO(^2\Pi)\,. \tag{48}$$

This excitation/dissociation scheme is an idealized (diabatic) representation. In reality the two electronic states are strongly coupled and must be treated simultaneously, especially when the pulse duration is longer than the mixing time of about 100 fs [130].

A rigorous dynamical investigation should involve coupled motion on both PES's. However, that would require two global PES's and the nonadiabatic coupling between them. This information is not available at the present time. Mahapatra et al. [130] recently determined two PES's and calculated the coupling element in a diabatic representation. However, they restricted their interest to the calculation of the absorption spectrum and the internal conversion and therefore considered only relatively small O–NO distances, too small for dissociation calculations. Global PES's for both states have been recently calculated by Reignier et al. [270]. However, the non-adiabatic coupling has not been determined by these authors.

In the absence of the necessary ingredients, classical trajectory [271] and quantum mechanical [35] calculations have been performed on a single, the lowest, adiabatic PES of the ground electronic state of NO_2. This PES does not have the same accuracy as those for HCO or HOCl. However, it provides a reasonably realistic picture of the interaction between O and NO in both the potential energy well and as they separate to form products. The dynamics of highly excited states of NO_2 is known to be irregular. The reasons are, first, the strong non-Born-Oppenheimer coupling between the two electronic states [272], second, the high density of vibrational states (due to the deep potential well and the three heavy atoms [271]), and third, the strong intra-mode coupling. Therefore, a state-by-state comparison with experimental data, as for HCO, is not meaningful and single-surface calculations are considered to be appropriate. For the same reason, the limited accuracy of the PES is also not problematic. Contour plots of the PES are depicted in Fig. 29. There is no barrier for dissociation and the TS is very loose as expected for two open-shell species. The structure between 110° and 120° originates from the conical intersection of the ground state with the first excited state.

The calculated state-specific rates $k(E)$, determined in Ref. 35 by harmonic inversion of an autocorrelation function as described in 4.1, are shown in Fig. 30. Because of the large density of states and the large DVR grid required, these calculations are extraordinarily demanding. There are two sets of calculated rates with different convergence criteria. The points above the dashed line represent states which are too broad to be considered as resonances. In contrast to HOCl, for example, none of the states can be assigned [see Fig. 5 (b)]. The general appearance of the fluctuations in the rates is similar to that for HNO (right-hand panel of Fig. 17). The fluctuations are about two orders of magnitude just above threshold and quickly decrease to about one order of magnitude as energy increases. The lowest calculated rates near threshold are of the same order as the rates obtained from the spectroscopic measurements of Abel et al. [137] (indicated by the narrow box at $E = 0$, see Fig. 30). Incidentally we note that the angular anisotropy parameter β, which to some extent reflects the lifetime of the complex, also fluctuates strongly as function of the excess energy [273].

Because of the severe congestion of the spectrum due to overlapping resonances, information about the time dependence of the dissociation process can be obtained only through time-resolved pump-probe experiments. They have been carried out by Ionov et al. [34] and by Kirmse et al. [35] as described in 3.1. Exciting a molecule by a pulse creates a wave packet,

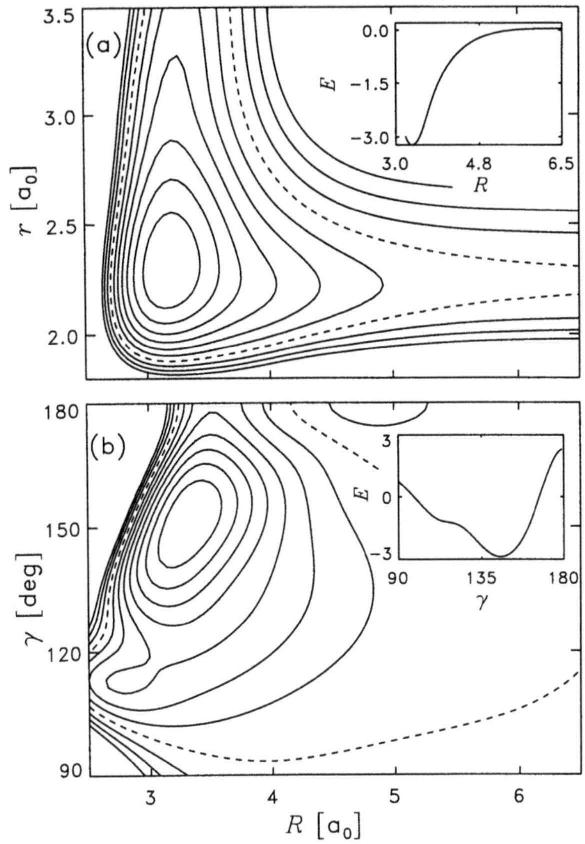

Figure 29: Two-dimensional contour plots of the NO_2 PES in Jacobi coordinates appropriate for O+NO. The insets show one-dimensional cuts (energy is in eV). Reprinted, with permission of the American Institute of Physics, from Ref. 271.

$$\Phi(t) = \sum_n A_n \Psi_n e^{-\mathrm{i}E_n t/\hbar}, \tag{49}$$

where Ψ_n are the resonance-state wave functions with complex energies E_n. The total probability for finding at time t the NO_2 molecule in any of the eigenstates can be estimated as

$$P(t) \approx \sum_n |A_n|^2 e^{-\Gamma_n t/\hbar}. \tag{50}$$

The coefficients $|A_n|^2$ are Franck-Condon-type factors; for an irregular system they can be taken as random numbers multiplied by a Gaussian function with spectral width ΔE centered around the mean excitation energy.

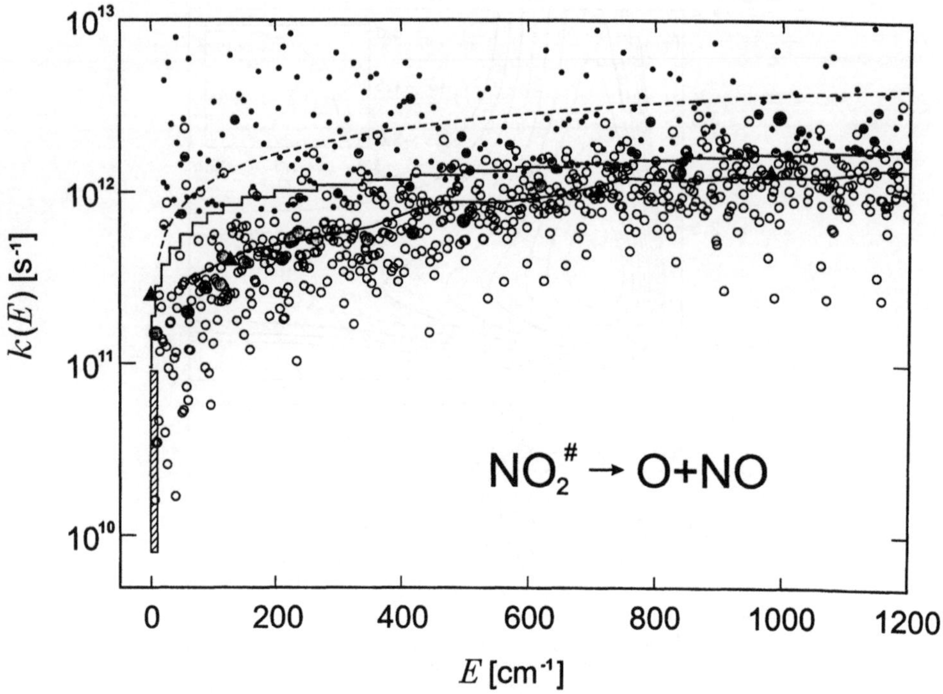

Figure 30: Calculated state-resolved dissociation rates for NO_2. The symbols indicate well converged (open circles) and less well converged (black dots) results. The smooth solid line indicates the quantum mechanical average (within windows of $\Delta E = 200$ cm^{-1}). The points above the dashed line correspond to lifetimes shorter than the ballistic time for ejecting one of the O atoms. The solid 'stepped' line is the SACM dissociation rate. The triangles represent the experimental average rates obtained by Kirmse et al. [35] and the shaded circles are the rates of Ionov et al. [34]. The hatched box at $E = 0$ shows the range of rates extracted from the energy-resolved spectroscopic experiment by Abel et al. [137]. The shaded boxes ($\Delta E = 200$ cm^{-1}) indicate the ranges of resonance states excited by the pump pulses with $\lambda_{pu} = 396$, 387, and 383 nm, respectively. Reprinted, with permission of the American Chemical Society, from Ref. 35.

According to Eq. (50), each resonance state contributes independently to the survival probability. Equation (50) can be shown to provide an appropriate approximation even for NO_2 [110] in which the resonances overlap. An interesting discussion of a more general decay function for overlapping resonances can be found in Ref. 274.

Kirmse et al. [35] measured the decay curves for four different excess energies, two close to threshold and two at around 1000 cm^{-1} above threshold (Figs. 30 and 31). The pulse duration is less than 1 ps and the spectral width is roughly 200 cm^{-1}. The experiments are performed in a molecular beam so that averaging over rotational states is marginal. The main ob-

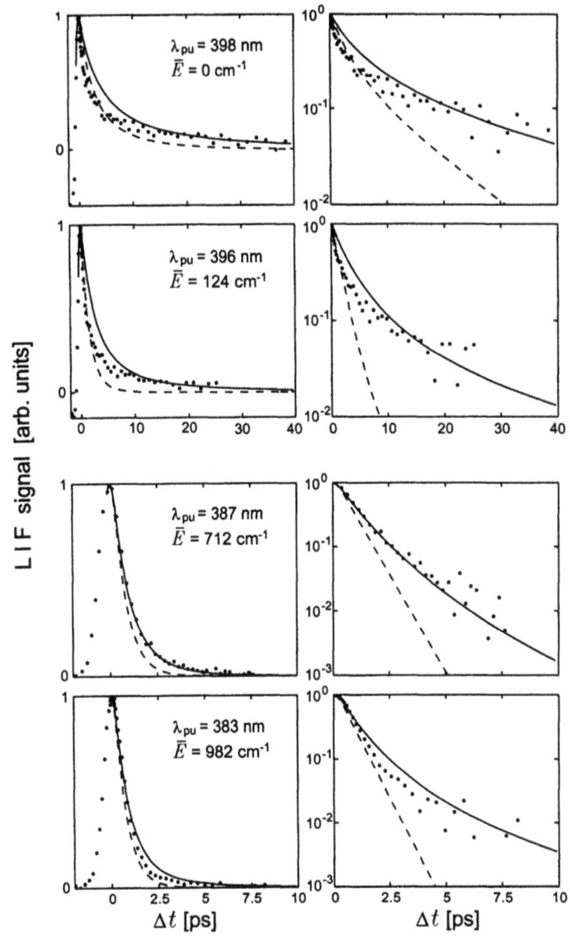

Figure 31: Time-dependent decay curves for the dissociation of NO_2 depicted on linear (left-hand panels) and on logarithmic (right-hand panels) scales. Dots: experimental results; solid lines: quantum mechanical decay curves obtained from the resonance poles; dashes: statistical decay curves obtained from the SACM rates. Note the different time scales. Reprinted, with permission of the American Chemical Society, from Ref. 35.

servation is that the decay curves are not exponential, with the deviation from a single-exponential decay being most pronounced at the lowest two excitation energies. The wave packet prepared by the photo-excitation is a superposition of many resonance states, Eq. (49); some of them decay very rapidly and others decay much more slowly. The tail of the decay curve at long times reflects the decay of the resonances having the smallest rates. It is the excitation of states with drastically different decay rates, which explains the non-exponential decay curves. We consider the non-exponential decay as a manifestation of the state-specific dissociation of NO_2 with fluctuations in rates up to two orders of magnitude. This result is not surprising in view of Fig. 30. It is, however, astonishing if one remembers the statistical character of NO_2 near threshold; all statistical models assume an exponential decay in such a case. The width of the distribution of rates is smaller at higher energies and therefore the decay is closer to a single exponential. Moreover, at higher energies dissociation is considerably faster. Thus, the signal intensity at long times is smaller, and the non-exponential tails at longer times are more difficult to see in the experiment.

The theoretical decay curves calculated by means of Eq. (50) and folded with the experimental resolution are also shown in Fig. 31 as well as the decay curves obtained from the SACM rates. Although the agreement is not perfect, the quantum mechanical decay curves do reproduce quite well the non-exponential behavior at intermediate and longer times. The slight curvature of the SACM decay curves in the logarithmic plots results from the strong energy dependence of $k_{\mathrm{SACM}}(E)$ near the threshold. The average rates extracted from the experimental decay curves for the four energies [35] agree reasonably well with the arithmetic average of the quantum mechanical rates (Fig. 30). They also agree with the rates measured by Ionov et al. [34], who considered the appearance of the NO product as function of time rather than the disappearance of NO_2. It should be noted that these authors fit their original data by a single exponential.

On the logarithmic scale it is clear that the SACM decay curves, on one hand, and the quantum mechanical and experimental curves, on the other, dramatically disagree at long times. In contrast, the initial fast decay, governed by the average rate, is well reproduced by the statistical theory. Note, that as much as 70% of the originally excited molecules decay within the initial time interval. The non-exponential behavior, which inherently cannot be described by statistical models, is the strongest when the survival probability drops below 30% or so. Thus, the gross decay of NO_2 is statistical in these experiments.

The decay of NO_2 at excess energies below $20\,\mathrm{cm}^{-1}$ has been investigated again by Bezel et al. [275] with the pump-probe technique using pulse widths of the order of 25 ps, which corresponds to a spectral width of $\leq 2\,\mathrm{cm}^{-1}$. The range of the measured rates depicted in Fig. 32 agrees with the range obtained from the spectroscopic experiments and is comparable to the range of rates calculated quantum mechanically (Fig. 30). Most striking are the pronounced structures as a function of energy, which *do not* correlate with the openings of new NO rotational product states. The large fluctuations of the quantum mechanical rates, especially near threshold, could provide one possible explanation for these structures. The rather

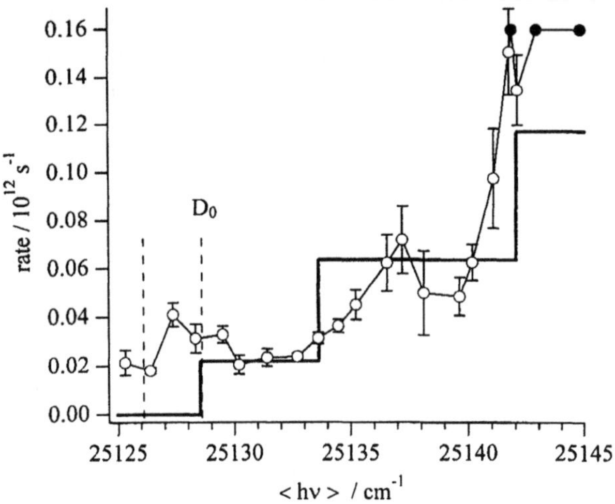

Figure 32: Dissociation rates of rotationally cold NO_2 as function of the mean excitation energy. Thick solid line: phase space theory (PST). The two vertical dashed lines are reaction thresholds for NO_2 starting in its lowest rotational level (i.e., D_0) and its first excited rotational level. Reprinted, with permission of the American Institute of Physics, from Ref. 275.

long pump pulse excites only very few states — in contrast to several tens of resonances, when a laser pulse of 1 ps duration is used. The initial wave packet contains only few resonances and its composition changes as the mean energy is increased. Because the widths of the individual resonance states differ by one order of magnitude and more, the average decay rate in a tiny energy interval can show noticeable changes with the mean energy. If the energy window covered by the wave packet significantly exceeds the spacing between adjacent states, the energy dependence of the average decay rate is smoother. At the same time, one cannot fully rule out another explanation of the features in Fig. 32 (cf. Ref. 275). Near the threshold, several closely spaced and asymptotically almost degenerate electronic states, each possibly supporting resonances, may be involved in the fragmentation. The dissociation of NO_2 at near-threshold energies is not yet fully understood, despite the many experimental and theoretical efforts.

NO_2 is one of the few molecules for which the dependence of the dissociation rate on the rotation of the excited complex has been investigated. In a double-resonance experiment, Bezel et al. [276] first excited NO_2 with an IR photon to a particular rovibrational state, $(1, 0, 1|J, K_a)$, and a second laser with fixed frequency promoted the prepared molecules into the continuum. Here, J is the total rotational angular momentum quantum number and K_a is the projection on the body-fixed frame. K_a is well defined at the bottom of the well, but ceases to be a good quantum number at energies near the dissociation threshold. The pump frequency is chosen so that only the initially tagged molecules can dissociate. By changing J in the intermediate state between 1 and 15, Bezel et al. could study the variation of $k(E, J)$ with J. Surprisingly, the rates for two different amounts of vibrational energies above threshold, $10\,\text{cm}^{-1}$ and $75\,\text{cm}^{-1}$, were found to be almost independent of J. The data are inconsistent with assuming either no coupling between the different rotational sub-levels or complete coupling. A simple model in which the coupling between neighboring K_a states is proportional to J^2 explains the data. The coupling is then attributed to Coriolis interaction. This picture is in accord with the findings of Grebenshchikov et al. [277] based on classical trajectories using the PES of Ref. 271. The more recent experimental results of Abel et al., obtained in time-resolved [278] as well as frequency-resolved [279] experiments, are consistent with the data of Bezel et al. [276].

6.3. Distribution of dissociation rates

In cases like D_2CO or NO_2 comparison with experimental data on a state-specific level are ruled out entirely and one has to retreat to more averaged quantities like the average dissociation rate, $\langle k \rangle$, and the distribution of rates, $Q(k)$. If the dynamics is ergodic — the basic assumption of all statistical theories — one can derive a simple expression for $Q(k)$, which had been established in nuclear physics in order to describe the neutron emission rates of heavy nuclei [280]. These concepts have since developed into the field of random matrix theory (RMT) and statistical spectroscopy [281–283] and have also found applications in the dissociation of energized molecules [121, 284–286].

In RMT one argues that the wave functions in the interaction part of the potential are so complicated that the matrix elements of coupling to the continuum have a Gaussian distribution with zero mean. The decay rates k can be shown to exhibit a χ^2-distribution of the form

$$Q_\nu(k) = \frac{\nu}{2\langle k \rangle} \left(\frac{\nu k}{2\langle k \rangle} \right)^{\nu/2-1} \exp\left(-\frac{\nu k}{2\langle k \rangle} \right) \Big/ \Gamma\left(\frac{\nu}{2} \right), \tag{51}$$

where ν is the number of 'degrees of freedom' and $\Gamma(\nu/2)$ in the denominator is the gamma function. The variance is $2/\nu$, that is, the distributions become narrower and narrower with increasing ν. Simultaneously, the position of the maximum of $Q_\nu(k)$ approaches $\langle k \rangle$. The number of degrees of freedom is usually given the physical interpretation of being the number of open decay channels. The number of quantum states at the TS with energies below the total energy E, i.e., the number of channels in the SACM seems to be a reasonable choice. Near the threshold the majority of states have rates much smaller than $\langle k \rangle$ and at higher energies, when more channels are energetically accessible, most rates have values near $\langle k \rangle$. All state-specific calculations and experiments show this general behavior. Generalizations of Eq. (51) have been discussed by Polik et al. [267]. Strictly speaking, the χ^2-distribution is only valid for non-overlapping resonances. A more general distribution for overlapping resonances was recently suggested by Fyodorov and Sommers [120]. The distribution of decay widths is still an active research field in physics; for a recent review see Ref. 120.

Figure 33 shows the distribution of all measured S_0 decay rates for D_2CO following excitation of the 4^1 band of the S_1 state [287]. The rates for all states with rotational quantum numbers J, K and M_J are taken into account. The experimental distribution is compared to the χ^2-distribution with $\nu = 6.1$ and good agreement is observed. The extrapolation to zero electric field yields an effective number of open channels of 3.8 which is connected to tunneling probabilities of a few open vibrational states at the transition state [287]. The RMT expression for the distribution of rates was later generalized by Hernandez et al. [116] to include angular momentum conservation and other dynamical symmetries.

The calculated dissociation rates for NO_2 in several energy intervals have also been compared to χ^2-distributions [35]. The effective number of degrees of freedom, which gives the best fit, has been found to be roughly a factor of two smaller than the number of open levels at the TS. The

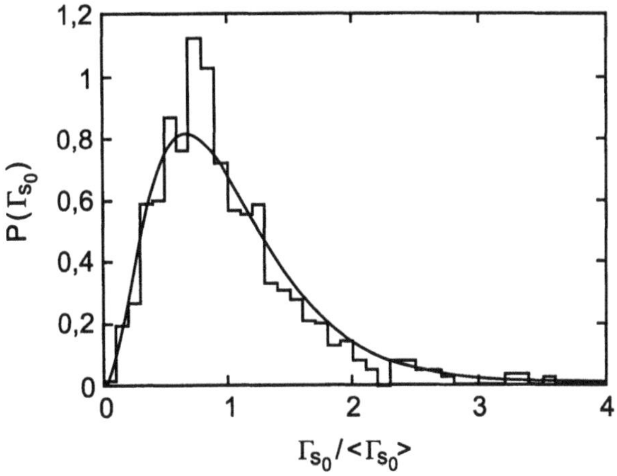

Figure 33: Distribution of decay widths measured for excitation of the 4^1 vibrational state of $D_2CO(S_1)$. The average width is $6.5 \times 10^{-4}\,\mathrm{cm}^{-1}$ corresponding to an average rate of $1.2 \times 10^8\,\mathrm{s}^{-1}$. The curved line is a χ^2-distribution with $\nu = 6.1$. Redrawn from Ref. 287.

same observation has also been made for the dissociations of HO_2 [60] and H_3^+ [62], two other molecules for which the ergodic limit appears to be valid. This observation reflects the fact that the transmission coefficient for a particular product channel is not immediately one, when this channel becomes energetically open. Incidentally we note that the distribution of mode-specific rates for HOCl cannot be fitted by a χ^2-distribution at all [67].

The measured time-resolved decay curves for NO_2 do not yield direct information about the distribution of rates. However, following Miller [288] and Lu and Hase [115] the χ^2-distribution can be transformed into a decay curve according to

$$P(t) = \int e^{-kt}\, Q(k)\, dk = \left(1 + \frac{2}{\nu}\langle k\rangle t\right)^{-\nu/2} . \tag{52}$$

$P(t)$ approaches an exponential in the limit $\nu \to \infty$. The comparison with the experimental decay curves for two excitation energies is shown in Fig. 34 for several values of ν. It is evident that in both cases $\nu = 2$ reproduces the experimental data well. Larger values of open channels suppress the smaller rates and therefore lead to curves, which fall off too quickly. However, $\nu = 2$ is significantly smaller than the predicted number of open adiabatic channels, which — for the PES of Ref. 271 — is about 8 for the average excess energy of 124 cm^{-1}.

6.4. 'Steps' predicted by RRKM theory

The RRKM expression for the dissociation rate, Eq. (42), predicts a stepwise increase of $k(E)$ with each opening of a new channel. The relative increase of the rate at a 'step', $\Delta k/k$, approximately scales as $1/N_{TS}$, i.e., it is the largest near threshold and successively diminishes as energy increases. The biggest 'step' is the one from $N_{TS} = 1$ to $N_{TS} = 2$. The energy intervals, at which successive 'steps' occur, are dictated by the frequency at the TS. Thus, molecules dissociating over a pronounced barrier would be best suited for observing 'steps', because the TS is tight and the successive 'steps' are separated by the TS-frequency of the order of 100 cm^{-1} or larger. Systems without exit channel barriers, on the other hand, often have very small TS-frequencies, $\sim 10\,cm^{-1}$, and the necessary energy resolution would be difficult to accomplish in order to see the 'steps'.

In 1992 Lovejoy et al. [289] claimed to have observed, for the first time, the 'steps' predicted by Eq. (42) in the unimolecular dissociation of triplet ketene, CH_2CO (Fig. 35). Triplet ketene dissociates over a barrier [33, 290] and the lowest TS-frequencies calculated [291] by ab initio methods prior to the experiment were in the range of 100–300 cm^{-1}. The observation

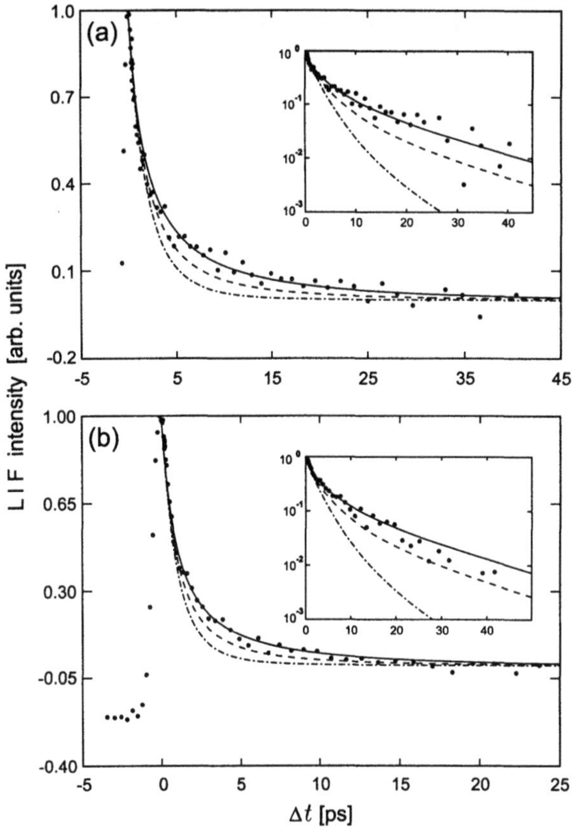

Figure 34: Test of the RMT prediction, Eq. (52), for the decay of NO_2 for $\lambda_{pu} = 398\,\text{nm}$ (a) and 396 nm (b). The dots represent the experimental decay curves. The number of open channels are $\nu = 2$ (solid lines), 4 (dashed lines) and 10 (dashed-dotted lines). The insets show the same results on a logarithmic scale. Reprinted, with permission of the American Chemical Society, from Ref. 35.

of structures by Lovejoy et al. in the plots of $k(E)$ versus E, was enthusiastically acclaimed by researchers working in the field of chemical kinetics [292, 293], because they were considered as proofs for the existence of quantized transition states in unimolecular dissociations. Shortly after these measurements, Ionov et al. [34, 294] reported to have found 'steps' in the dissociation rate of NO_2, which they also attributed to the quantized TS according to RRKM theory. In Fig. 30 the experimental data of Ionov et al. are shown on a logarithmic scale and the 'steps' are not so obvious. A linear scale reveals them more clearly; there are two pronounced structures, near $100\,\text{cm}^{-1}$ and $200\,\text{cm}^{-1}$, respectively. Note that NO_2 dissociates through a loose TS and therefore these observations are

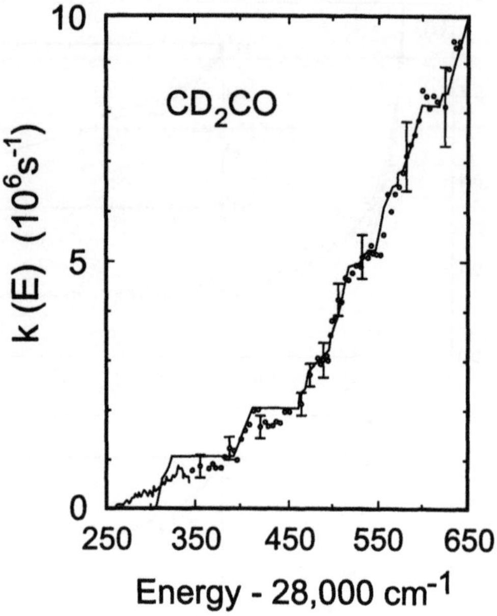

Figure 35: Measured dissociation rate constant $k(E)$ for CD_2CO. The solid line is a fit based on RRKM theory including tunneling corrections. Redrawn from Ref. 33.

quite unexpected. Structures were also observed in the dissociation rate constant of triplet acetaldehyde, CH_3CHO [36]. They were also associated with the quantum-level structure of the TS.

The idea of a quantized TS is certainly very appealing, from a conceptional as well as a practical point of view. If the structures in $k(E)$ would correspond to the TS energy levels, they could be exploited to gain information about the PES in the region of the TS. There is, however, one fundamental difficulty in relating the observed 'steps' to the quantized transition state. In *all* quantum mechanical calculations performed up to now as well as in the few state-resolved experiments (D_2CO, CH_3O, HCO, and DCO) the decay rates are found to fluctuate over several orders of magnitude. These fluctuations are especially pronounced and the distribution of rates is the widest near the dissociation threshold – exactly where the 'step' in the statistical rate is expected to be the largest. The increase of the statistical (i.e., average) rate by a factor of two is much smaller than the breadth of the distribution of rates – even if the narrowest resonances are not taken into account. Thus, the variation of the quantum mechanical rates in a small energy interval exceeds greatly the energy variation of $k(E)$ as predicted by RRKM theory at a 'step'. Moreover, opening of the next adiabatic channel has little effect on the quantum distribution

— even at threshold (see, for example, Fig. 30). One might argue that the distribution of rates shrinks with increasing density of states and large molecules are more likely to feature 'steps' than small ones. However, while the absolute values of the rates do decrease with increasing ρ, the threshold distribution of *relative* rates, i.e., measured in units of the average rate, is qualitatively similar to the Porter-Thomas distribution, irrespective of whether one considers HNO, NO_2 or CH_2CO. For all systems, this distribution predicts existence of many states whose decay rate deviates substantially from the average.

The existence of 'steps' also can be questioned by reviewing the main assumptions made in RRKM theory. First, RRKM is essentially a classical theory, i.e., the dynamics, especially motion along the dissociation coordinate, is treated by classical mechanics (see also 8.1). The 'quantization' of the numerator in Eq. (42) — introduced by quantizing the motion transverse to the reaction path — is a supplemental ad hoc modification; it is not rigorously derived from the original Schrödinger equation. In particular, it completely neglects the state-specific nature of the coupling of the resonance states to the continuum. Second, the energy levels at the TS are defined in the adiabatic approximation, discussed in 2.2, in which the 'slow' dissociation mode is decoupled from the 'fast' vibrational modes. In a more accurate description also the non-adiabatic matrix elements, which couple the different adiabatic channels, have to be taken into account. They are usually not small at the TS where the potential 'perpendicular' to the reaction path and likewise the adiabatic vibrational wave functions change considerably as the reaction coordinate varies. Inclusion of non-adiabatic coupling would tend to smear out the 'steps'. Third, tunneling through the one-dimensional adiabatic potential curves also tends to round-off the step-like structures predicted by the RRKM expression as was convincingly demonstrated by model calculations for ketene by Gezelter and Miller [295].

As a result, the attempts to quantify the observed structures in $k(E)$ using RRKM lead to inconsistencies. In NO_2, the observed 'steps' are separated by $100\,cm^{-1}$ – which should be considered as the lowest TS-frequency. In contrast, the lowest adiabatic potential curves obtained from the ab initio PES of Grebenshchikov et al. [271] have no or only very small barriers, i.e., the TS-frequency is certainly much smaller than $100\,cm^{-1}$. Discrepancies between the RRKM predictions of the TS-frequency and ab initio calculations were also found in triplet ketene, for which the analysis of the step-like structures in the observed energy profile of $k(E)$ was completed by Kim et al. [33]. In this molecule, tunneling through one-dimensional adiabatic potential barriers plays an important role and must be included in the statistical treatment. The barrier transparency, semiclassically related to the imaginary frequency at the barrier top, becomes an adjustable parameter. The RRKM modeling gave empirical barrier frequencies in the range of $(100 \pm 40)\,\mathrm{i\,cm^{-1}}$, which were at least a factor of three lower than the ab initio results available at that time [291]. The statistical model predicts a flat broad barrier with low transparency in order to keep the 'steps' from smearing out by tunneling contributions [295]. More accurate electronic structure calculations by King et al. [296] did not solve this puzzle; the newly calculated barrier frequency of $321\,\mathrm{i\,cm^{-1}}$ is still much too large compared to the RRKM estimate. The calculated barrier is narrow and therefore tunneling is likely to wash out any step-like behavior. The same observation was made for acetaldehyde [297]. Using two different electronic structure theories, barrier frequencies of $355\,\mathrm{i\,cm^{-1}}$ and $516\,\mathrm{i\,cm^{-1}}$ were found. Although both values are different, they deviate greatly from the value inferred from the RRKM model, which is only $60\,\mathrm{i\,cm^{-1}}$ [36]. This lead King et al. [297] to conclude: "Thus, while the experimental steps in the dissociation rate of acetaldehyde (as well as ketene) have yet to be satisfactorily explained, proven ab initio methods provide strong evidence that simple RRKM fits to the $k(E)$ profile provide misleading vibrational frequencies of the transition state on the corresponding triplet potential energy surface."

The measured dissociation rates $k(E)$ for CH_2CO, NO_2, and CH_3CHO undoubtedly exhibit structures, whose origin is not clear even one decade after their first observation. A possible explanation may be the large fluctuations of the state-specific dissociation rates. In experiments one generates a superposition of states, i.e. a wave packet, rather than a single state. For example, in the experiment of Ionov et al. [34, 294] the spectral width is about $250\,\mathrm{cm^{-1}}$, which means that many resonance states are simultaneously excited. The measured rate then approximately equals the average of the individual rates contained in the spectral windows. In view of the wide fluctuations it appears not implausible that some slight structures may survive this averaging process. The calculated average rate for NO_2 (Fig. 30) exhibits some undulatory behavior for the spectral width of $200\,\mathrm{cm^{-1}}$. However, these structures are not correlated with the quantized TS as the comparison with the SACM rate readily proves. Narrower energy windows would give rise to more and more pronounced structures. In conclusion, the yet not satisfactorily explained structures in the energy-dependent dissociation rate for molecules like NO_2 or ketene should stimulate theorists to perform rigorous dynamics calculations on global PESs, despite the fact that such calculations will be highly demanding.

7. PRODUCT STATE DISTRIBUTIONS

The distribution with which the individual quantum states of the products are populated contains additional information about the fragmentation process, especially about the dynamics beyond the transition state [20]. The list of final state properties accessible to modern detection methods is exhaustive. Most commonly, the state of the fragments is characterized with scalar quantities, such as the distributions of vibrational and rotational energies. The vibrational state distributions reflect the variation in the bond length of the fragment molecule along the dissociation path, and rotational distributions are related to the changes of the bond angle during the fragmentation. The energy partitioning in the fragments might be important for subsequent reactions, because often the product species are created in highly excited states. If the electronic angular momenta and spins of the fragments are nonzero, the rotational product states are further split into fine structure components. In this case, additional information about the electronic states involved in the fragmentation can be obtained by measuring the relative populations of the Λ doublet and/or spin orbit states, as was done, for example, for H_2O [298], NO_2 [299], and NH_3 [300]. Furthermore, if a polyatomic molecule breaks into two fragments with at least two atoms each, like singlet H_2CO, one can consider more elaborate quantum-state correlations, which describe the joint probability for the two fragments to be in well defined rotational states simultaneously [301] (see also recent work on ketene in Ref. 302).

Spatially resolved information about the reaction is provided by so-called vector correlations. The electric field vector $\mathbf{E}$ of the polarized laser light defines a natural space direction, with respect to which the vector quantities can be measured, such as the transition dipole moment $\boldsymbol{\mu}$, the fragment recoil velocity $\mathbf{v}$ or the fragment angular momentum $\mathbf{j}$. Well known is the correlation between $\mathbf{E}$ and $\mathbf{v}$, which gives rise to anisotropic angular distributions of photo-fragments [303, 304]. This correlation, described through a single anisotropy parameter β, characterizes the type of electronic transition (i.e., parallel or perpendicular), the electronic symmetry of the metastable state and, indirectly, its lifetime (cf. Refs. 303 and 305). Recently, the β parameter, strongly fluctuating as function of the excitation energy, was accurately measured for NO_2 [273]. Further correlations exist between $\boldsymbol{\mu}$ and $\mathbf{j}$ as well as between $\mathbf{v}$ and $\mathbf{j}$ [20].

In this section, we shall focus exclusively on the scalar properties of the fragments and consider the vibrational and rotational product state distributions (PSD's) following the dissociations of HCO, NO_2, and H_2CO discussed in Sects. 5 and 6. An in-depth introduction to the vast and fascinating field of product state analysis can be found in Ref. 20 (Chapters 9, 10, and 11). Recently, the PSD's of several representative groups of molecules were reviewed in Ref. 306.

In general, what has been said about the rates does also apply to the

PSD's. They are reflections of the underlying wave functions, their nodal structures, *and* the dynamics in the exit channel. As outlined in detail in Ref. 20 (Chapters 9 and 10), in many cases the wave function at the TS defines the starting conditions for the final step of the fragmentation process: If the system shows mode specificity, the PSD's also will show qualitative behaviors which are typical for excitation of particular modes. However, if the dissociation rates show statistical state-specific behavior, it *does not* necessarily follow that the PSD's have a statistical dependence on the quantum numbers of the fragments. An illuminating example is the dissociation of H_2CO to be discussed in 7.3.

7.1. General considerations

The theoretical description of PSD's is based on the notion of a partial cross section, $\sigma_\alpha(E)$, for the production of a particular quantum state α of the fragments [20, 307, 308]. The product distribution is given by the relative populations of open channels — it is defined as a ratio of $\sigma_\alpha(E)$ and the total cross section $\sigma(E)$, Eq. (13) [20]. In what follows we shall consider only vibrational and rotational state distributions, so that the channel index α is assumed to comprise only two quantum numbers, j and n, corresponding to the rotation and vibration of the diatomic fragment, respectively. In this case, the partial cross section for a particular rovibrational channel is $\sigma_{jn}(E)$. The cross section for a given vibrational channel equals $\sigma_n(E) = \Sigma_j\, \sigma_{jn}(E)$ and the total cross section is given by $\sigma(E) = \Sigma_{jn}\, \sigma_{jn}(E)$. Accordingly, the product rotational distribution in a given vibrational manifold is defined as

$$P_j^{(n)}(E) = \sigma_{jn}(E)/\sigma_n(E)\,, \tag{53}$$

while the product vibrational distribution is given by

$$P_n(E) = \sigma_n(E)/\sigma(E)\,. \tag{54}$$

In the time-independent scattering formalism, the partial cross section is proportional to the square of the overlap between the scattering wave function $\Psi^{(jn)}$ in channel $\{j, n\}$ [cf. Eq. (17)] and the initial molecular state Ψ_0. The distributions $P_n(E)$ and $P_j^{(n)}(E)$ can be estimated directly from the solutions of the time-independent Schrödinger equation described in 4.2. In general, one obtains different PSD's for different initial states Ψ_0 of the molecule.

In the time-dependent approach, a convenient method for calculating partial cross sections was suggested by Balint-Kurti et al. [308]. In the asymptotic region, at a fixed value of the reaction coordinate $R = R_\infty$, the evolving wave packet $\Phi(R_\infty, r, \gamma; t)$ is projected onto the orthonormal basis of the channel wave functions $\phi_{jn}(r, \gamma)$, which we introduced in Eq. (18), i.e.,

$$\Phi(R_\infty, r, \gamma; t) = \sum_{jn} C_{jn}(R_\infty, t)\phi_{jn}(r, \gamma)\,. \tag{55}$$

In the next step, the time-dependent coefficients $C_{jn}(R_\infty, t)$, calculated by $\langle\phi_{jn}(r, \gamma)|\Phi(R_\infty, r, \gamma; t)\rangle$, are Fourier transformed to give the partial cross sections [308] according to

$$\sigma_{jn}(E) \propto \left|\frac{1}{2\pi}\int_0^\infty dt\, e^{\mathrm{i}Et/\hbar}\, C_{jn}(R_\infty, t)\right|^2. \tag{56}$$

Again, different PSD's result from different initial wave packets. Equation (56) can be used not only in wave packet calculations, but also, after simple modifications, in any other iterative scheme. One has to ensure, however, that the 'observation point' $R = R_\infty$ lies outside the interaction region *and* outside the region of non-zero absorbing potential.

Each resonance state is uniquely characterized by an individual PSD. Suppose that the wave packet in Eq. (55) consists of only one state,

$$\Phi(R, r, \gamma; t) = \Psi_{\mathrm{res}}(R, r, \gamma)\, e^{-\mathrm{i}E_0 t/\hbar}\, e^{-\Gamma t/2\hbar}\,, \tag{57}$$

where $\Psi_{\mathrm{res}}(R, r, \gamma)$ is the (unnormalized) coordinate part of the resonance wave function. Then the Fourier integral in Eq. (56) can be easily taken to give the partial cross section,

$$\sigma_{jn}^{\mathrm{res}}(E) \sim \frac{|\langle\phi_{jn}(r, \gamma)|\Psi_{\mathrm{res}}(R_\infty, r, \gamma)\rangle|^2}{\|\Psi_{\mathrm{res}}(R_\infty, r, \gamma)\|}\, \frac{\Gamma/2}{(E - E_0)^2 + (\Gamma/2)^2}\,, \tag{58}$$

where $\|\cdots\|$ indicates the norm of the wave function. Note, that in every channel $\sigma_{jn}^{\mathrm{res}}(E)$ has the same Lorentzian profile, as the total cross section [cf. Eq. (4)]; only the intensities of the lines, proportional to the projections of Ψ_{res} on the channel wave functions, are different for different channels. For this reason, the PSD,

$$P_{jn}^{\text{res}}(E) = \frac{|\langle \phi_{jn}(r,\gamma)|\Psi_{\text{res}}(R_\infty, r, \gamma)\rangle|^2}{||\Psi_{\text{res}}(R_\infty, r, \gamma)||}, \tag{59}$$

remains constant as the energy is varied across the resonance. Equation (59) gives a practical way of evaluating the product distributions from the known resonance wave functions. The analysis of PSD's of individual states is beneficial for non-overlapping assignable resonances — like those for HCO. If the widths exceed the level spacings, one has to determine the distributions, which then depend on the energy, by the more general approaches described above, i.e., from the full solutions of the time-dependent or time-independent Schrödinger equations.

7.2. Vibrational state distributions

HCO is a mode-specific assignable system, and resonance states in each progression yield distinct vibrational state distributions. Let us consider the pure CO stretching progression $(0, v_2, 0)$ already discussed in 5.1. The PSD's for this case can be explained in terms of a simple dynamical model and their evolution within the progression can be predicted — at least qualitatively. Figure 36 shows the vibrational distributions for the resonances with $v_2 = 8$–11 [49]. They are calculated using the scattering wave functions, Eq. (17), at energies E_0 corresponding to the center of each resonance (remember, that according to Eqs. (58) and (59) the PSD within an

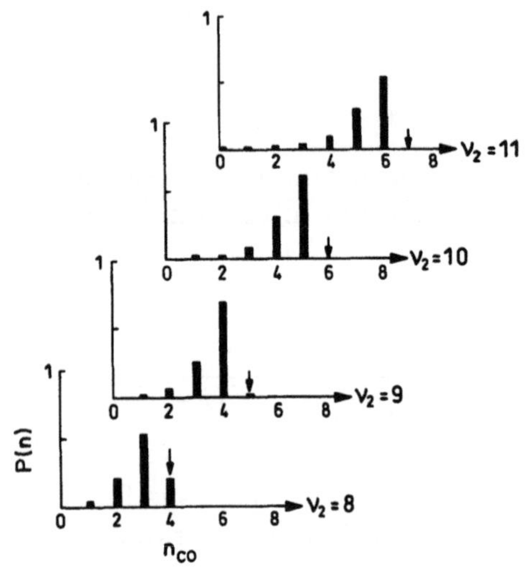

Figure 36: Calculated vibrational state distributions for CO following the decay of the $(0, v_2, 0)$ pure CO stretching resonances of HCO. The arrows indicate the highest accessible vibrational product state. Reproduced, with permission of the American Institute of Physics, from Ref. 49.

isolated resonance is energy-independent). All distributions peak at high fragment quantum numbers n_{CO}, fall off very steeply to higher vibrational states, and decline more gradually to the lower end. The pronounced peak occurs close to the energetic cutoff $n_{max}(E_0)$, i.e., the highest CO state, which can be populated at E_0. The same general behavior is also observed for the lower resonance states with $v_2 < 8$ [49].

In Ref. 49 these distributions were explained in terms of a vibrationally non-adiabatic process in which energy from the originally excited CO mode is transferred step by step to the other two modes, including the dissociation bond. The model is very similar to the non-adiabatic 'tier' mechanism for the narrow Feshbach resonance (7,0,0) in HOCl, discussed in 2.2. In the present case, 4–5 CO vibrational quanta are required to be swapped before the first energetically accessible CO vibrational state is reached. If HCO then quickly dissociates without being hindered by a barrier on the respective adiabatic curve, the product state with $n_{CO} = n_{max}(E_0)$ is populated with the maximum probability. This happens with moderately excited resonance states with $v_2 < 8$ (not shown in Fig. 36). If, however, the molecule is trapped behind the barrier in the adiabatic state $n_{CO} = n_{max}(E_0)$, the transfer of another vibrational quantum might be more effective, and the state $n_{CO} = n_{max}(E_0) - 1$ dominates the PSD. The latter scenario is apparently valid for all distributions shown in Fig. 36. The probability for filling the lower vibrational states gradually decreases with $|n_{max}(E_0) - n_{CO}|$. This is in accord with the energy-gap model developed for the predissociation of weakly bound van der Waals complexes [100]: The larger the translational energy in the dissociation coordinate, the smaller is the probability. Unfortunately, the experimental information on the vibrational state distributions in the photolysis of HCO is rather sparse [244] so that comparison with theory is very limited. Nevertheless, for the few resonances for which a comparison is possible, the agreement is acceptable [49].

Because of the evident mode specificity in the HCO dissociation, classical calculations with the initial conditions for trajectories selected from a micro-canonical, uniform distribution over the whole available phase space (see Sect. 8.1) cannot describe the PSD's of individual resonances in a satisfactory way. However, without presenting results, we mention that the classical distributions at least qualitatively reproduce the high degree of inversion. The statistical models, on the other hand, fail severely in reproducing the pronounced propensity for filling the highest accessible vibrational states. All statistical models predict that the populations of the various vibrational channels are proportional to the number of energetically open rotational levels within each vibrational manifold. Therefore, regardless of whether we count the states of the free diatom like in phase space theory (PST) [309] or at finite distances as in the SACM, the vibra-

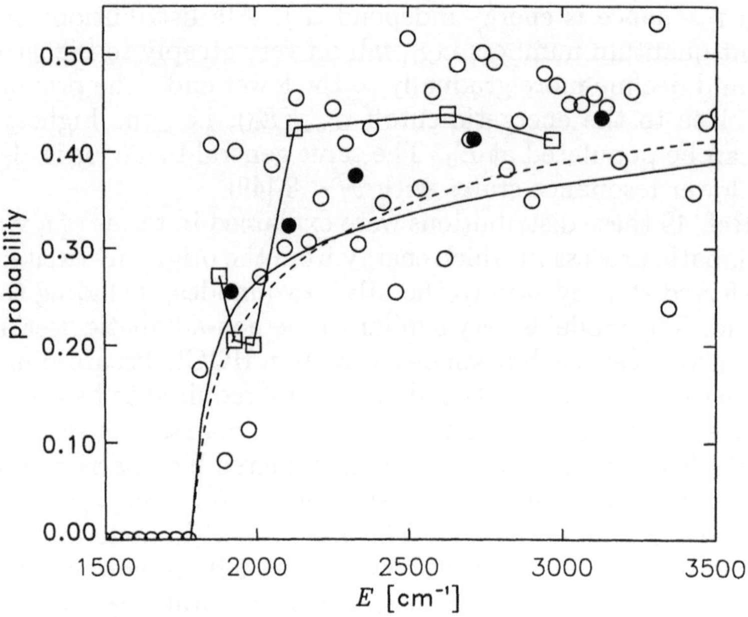

Figure 37: The probability of populating the vibrational state $n = 1$ of NO as a function of energy in excess of the first dissociation threshold in the dissociation of NO_2. Open circles: quantum mechanical results; filled circles: classical calculations; solid line: SACM; dashed line: PST; open squares: experimental results from Ref. 133.

tional ground state $n_{CO} = 0$ has the maximum probability, in complete disagreement with the quantal results.

Dissociation of nitrogen dioxide is much more statistical and this is reflected in the vibrational PSD's as well. Since the resonance states are unassignable [cf. Fig. 5(b)] and, moreover, the neighboring resonances overlap, the state-resolved PSD's like those for HCO in Fig. 36 cannot be plotted for NO_2. Instead, one can analyze energy-dependent PSD's, Eqs. (54) and (56), for a given initial wave packet, which resembles the molecular state immediately after the laser excitation and encompasses many overlapping resonances. In Fig. 37 we show the result of such an analysis in a relatively narrow energy window just above the dissociation threshold for $n_{NO} = 1$. Plotted is the probability for filling the first excited vibrational state of NO, $P_{n_{NO}}(E)$ [39]. It steeply rises at the threshold and fluctuates just like the dissociation rates. The average probability very quickly approaches the statistical limit of 0.5, i.e., the ground and the first excited fragment states are populated with equal weights. The SACM and PST predictions go through the quantum mechanical points as expected for a mainly statistical system. Both statistical theories are close to each other because the TS for the reaction is loose — the potential along the minimum energy path is barrier-less. In agreement with the statistical and quantum mechanical predictions is the probability from classical trajectory calculations. Since the classical dynamics is strongly chaotic [271], the resulting distribution is insensitive to the particular choice of the initial conditions in the inner part of the potential: IVR is faster than the dissociation. Because of the violation of conservation of zero-point energy, the classical probabilities are nonzero already below the quantum threshold.

Quantum mechanical, classical and statistical probabilities agree, on average, reasonably well with the experimental results [133] shown in Fig. 37 (vibrational distributions of NO were also measured by Harrison et al. [310]). In the experiment a high population of the state $n_{NO} = 1$ is found already 100 cm^{-1} above its threshold. Moreover, the measured probabilities show some indications of fluctuations. Because of the limited number of data points, the inevitable incoherent averaging over several overall rotational states of NO_2 and the averaging over the various possible electronic states of the O and NO products, these fluctuations are less pronounced than in the quantum mechanical calculations on a single adiabatic PES and for $J = 0$.

The fluctuations of PSD's, like those in Fig. 37, are not limited to irregular unassignable molecules. Actually, had the same probability been plotted for HCO as a function of energy, one would also find fluctuations. Moreover, their range would be much broader, covering the whole interval [0,1]. In fact, a *narrow* distribution about the average and an average probability close to the statistical limit of 0.5 are the hallmarks of the statistical dissociation of NO_2.

7.3. Rotational state distributions

The product state distributions cannot be predicted solely on the basis of the dynamical patterns in the inner part of the potential. This means, in particular, that a molecule, whose internal dynamics is statistical state-specific, might, nevertheless, dissociate into fragments with non-statistical rotational distributions. Actually, rotational PSD's reflect the angular shape of the resonance wave functions *at* the TS, rather than the wave function in the well region, and the dynamics in the exit channel beyond the TS, i.e., the torque which the separating atom exerts on the diatom (mapping of the transition-state wave function, Ref. 20, Chapter 10, and Ref. 311). Three generic situations are considered below. In the case of a very high barrier in the fragment channel, the diatom is highly excited, because the torque is very likely strong and the released energy is large. Dissociation of formaldehyde into H_2 and CO is a prototype. The torque is significantly weaker in systems with a low barrier, which makes rotational distributions much colder — as in the dissociation of HCO. Finally, if there is no real potential barrier, like for NO_2, the torque beyond the TS is small, and the overall shape of the rotational distribution mainly reflects the bending excitation of the molecule at the TS.

Rotational state distributions of the CO diatom in the dissociation of formaldehyde have been studied by Moore and coworkers using LIF spectroscopy [312]. As discussed in 6.1, the molecule is first promoted to the excited electronic state S_1, which is strongly coupled to the ground state S_0, and the dissociation finally takes place on the S_0 PES. The experimental results are shown in Fig. 38. PSD's are measured for different initial rovibration states in the S_1 electronic state and for given vibrational states of CO, $n_{\rm CO} = 0$ and 1. All measured distributions are bell-shaped and peak at high values of $j_{\rm CO}$, showing a weak dependence on the initial state. They are clearly non-statistical. These regular distributions, also observed for the isotopically substituted species HDCO and D_2CO (see Fig. 39), contrast sharply with the erratic fluctuations of the dissociation rates. The strongly chaotic dynamics of highly vibrationally excited H_2CO has little effect on the rotational PSD's. The physical reason for this is the high barrier which separates formaldehyde from the $H_2 + CO$ exit valley [312–315]. The energy pumped into H_2CO merely suffices to overcome the barrier, beyond which the recoiling products slide down the

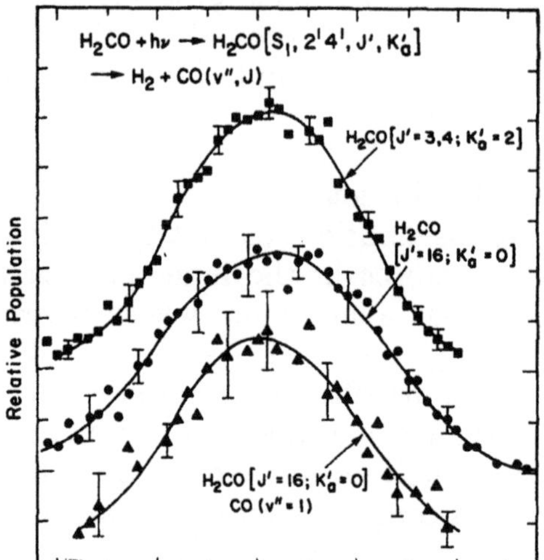

Figure 38: Experimental CO photo-fragment rotational distributions following the dissociation of H_2CO. The two upper traces are for $CO(n = 0)$, while the bottom trace is for $CO(n = 1)$. Reproduced, with permission of the American Institute of Physics, from Ref. 312.

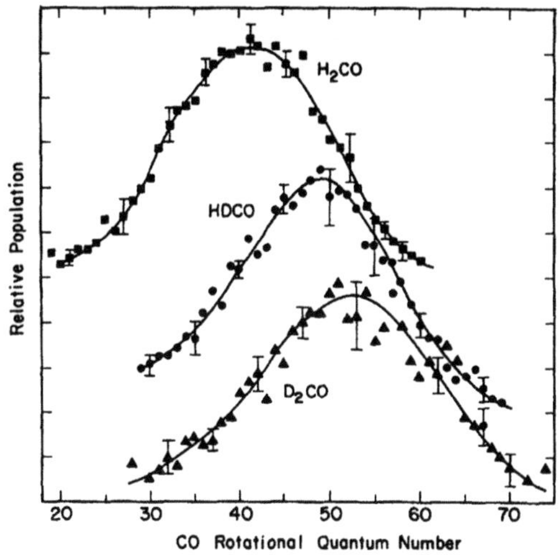

Figure 39: Effect of isotopic substitution on CO(n = 0) rotational distributions. The top trace is for H_2CO from the top trace in Fig. 38. The middle trace is for HDCO photolysis at ~ 29 515 cm^{-1} and the bottom trace is for D_2CO photolyzed at ~ 29 545 cm^{-1}. Reproduced, with permission of the American Institute of Physics, from Ref. 312.

slope of the steep exit channel potential. The dissociation from the TS is, therefore, a nearly direct process. The authors of Ref. 312 conclude: "After the molecule passes through the energy saddle point on its path to dissociation, the newly forming photo-fragments experience such a strong repulsion that they fly apart very rapidly, on a timescale shorter than the timescale of vibrational or rotational energy transfers. Therefore, energy is *not* statistically redistributed."

A simple but quantitatively accurate model of the decay through the narrow TS was suggested by Schinke [314]. The model is based on the observation that near the barrier top almost no energy is left in the degrees of freedom transverse to the reaction path — the transverse modes can perform only zero-point vibrations. Thus, the wave function of H_2–CO along the bending coordinate γ at the TS can be assumed to be a simple Gaussian function, despite the fact that in the inner part of the potential the eigenstates have complex unassignable nodal structures. Moreover, the steepness of the exit channel justifies the use of the quantum mechanical sudden approximation for the rotational dynamics [20]. The rotational PSD is assumed to be entirely defined through the *dynamical mapping* [20] of the angular state at the TS, $|\Psi_{TS}(\gamma)|^2$, onto spherical harmonics describing the free rotor fragment states. This mapping reflects the anisotropy of the interaction potential in the H_2 + CO channel, which is known from electronic structure calculations [316]. Since $|\Psi_{TS}(\gamma)|^2$ is a Gaussian, the

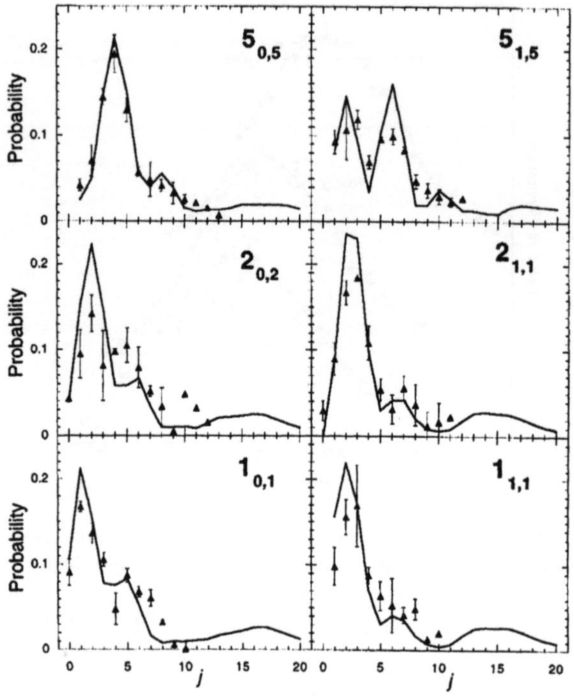

Figure 40: Theoretical (solid lines) and experimental (symbols, Ref. 244) $n = 2$ CO rotational distributions for selected rotational levels of the resonance (0,7,0) of HCO with total angular momentum $J = 1$, 2, and 5. Reproduced, with permission of the American Institute of Physics, from Ref. 52.

PSD turns out to be a Gaussian-like function of j_{CO} too. The model applies equally well to the isotopically substituted formaldehydes (Fig. 39), for which the observed shift of PSD's to higher j_{CO} values is proportional to the square root of the reduced mass of the molecule. Chang et al. studied the decay of formaldehyde using classical trajectories on a semiempirical six-dimensional PES [315] starting the trajectories at the TS. Since the final step of the reaction is very fast, classical dynamics produce PSD's which agree well with the experiment.

Signatures of the wave functions at the TS are also clearly visible in the rotational distributions of CO in the dissociation of HCO [52,244,317]. The barrier for H + CO production is significantly lower than in H_2CO, so that the torque exerted on CO is not strong, but not negligible either. Since the resonance states of HCO are assignable and non-overlapping, one can consider PSD's emerging from individual states [cf. Eq. (59)]. Figure 40 shows — for different HCO rotational states — rotational state distributions of CO($n = 2$) following the decay of the resonance (0,7,0) as calculated by Yang and Gray [52]. Similar results have been obtained by Keller and

Schinke [317]. The rotational levels of the parent molecule are labeled by J_{K_a,K_c} with K_a and K$_c$ being projections of J onto the a and c molecular axis, respectively. All distributions for the various HCO rotational states have essentially the same general behavior, a main peak at small values of j, a minimum at intermediate j states, and a lesser pronounced maximum around $j \approx 15$. The clearly discernible bimodal behavior is a direct consequence of the angular dependence of the TS wave function, which has a node in the vicinity of the barrier [317]. The main maximum is overlayed by oscillations, which appear to be random. However, they can be unambiguously explained by a model based on Franck-Condon mapping and intrinsically reflect the dependence of the corresponding wave functions on the overall rotational state [318]. They vary with the total angular momentum quantum number J and its projections K_a and K_c. Thus, for HCO, both the gross structures and the finer details are explainable and therefore predictable in terms of the structure of the underlying resonance wave function. The PSD's for resonances belonging to the same progression are qualitatively similar, but may differ from the distributions for states

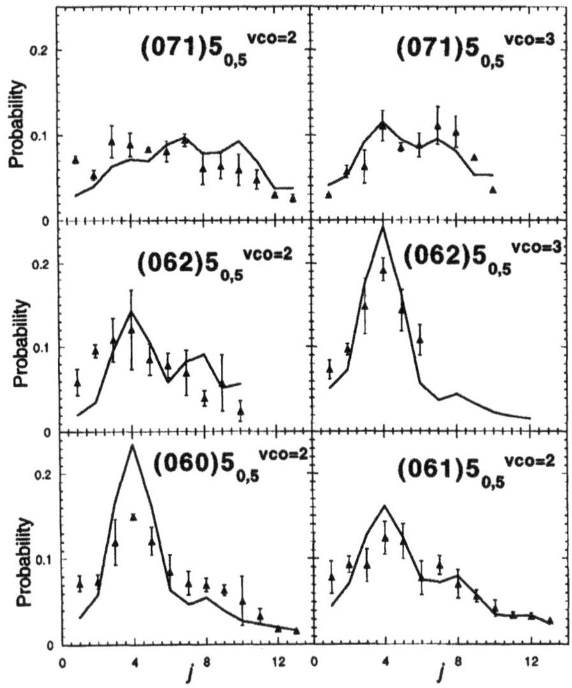

Figure 41: Theoretical (solid lines) and experimental (symbols, Ref. 244) CO rotational distributions for resonance states from various progressions with rotational excitation $5_{0,5}$. Reproduced, with permission of the American Institute of Physics, from Ref. 52.

from other progressions. Examples for the resonance states (0,6,0), (0,6,1), (0,6,2), and (0,7,1) are given in Fig. 41. There is an obvious similarity between the distributions for $(0,6,0)_{0,5}$ and $(0,7,0)_{0,5}$ in Figs. 40 and 41. At the same time, the distributions in the adjacent progressions show different population patterns. This is again a manifestation of mode specificity: The shapes of resonance wave functions at the TS are different for different progressions.

Also shown in Figs. 40 and 41 are experimental data measured by Neyer et al. [244]. The agreement is excellent and the basic theoretical predictions are confirmed. In contrast to the quantum mechanical calculations, all statistical models predict — for $J = 0$ — constant distributions, which abruptly vanish at a specific maximum value of j and thus fail to describe the details of the quantum mechanical and experimental distributions. The same is also true for the classical distributions.

The population of rotational states of NO in the dissociation of NO_2 proceeds in a different way. Nitrogen dioxide lacks both the assignability of HCO and the transition state barrier of H_2CO. In fact, the potential along the reaction path and the lowest adiabatic curves are barrier-less, the TS is very loose and can be detected only with the variational RRKM [271]. As a result, the torque exerted on NO is small, whereas the angular behavior of the wave functions in the TS region is as complicated as in the interaction part of the potential. The 'initial conditions' for the final decay are nearly random. This is reflected in the NO rotational state distributions shown in the upper panels of Fig. 42 for two energies. The quantum mechanical results are obtained using the wave packet propagation as described in 7.1. The distributions considerably fluctuate vs. j_{NO}, and the fluctuations appear to be much less predictable than for HCO. This is partly because the unassignable wave functions of NO_2 have an irregular angular dependence at the TS. Additionally, the non-adiabatic interactions between rotational channels do not vanish immediately beyond the variational TS and cause further, more random like, redistribution of the populations even at larger separations. For this reason, simple Franck-Condon mapping models do not explain the observed distributions. Because of the post-TS interactions, for low excess energies the distribution is spread over the full range, $[0, j_{\max}(E)]$, of the accessible rotational states [Fig. 42(a)]. With increasing energy, the dissociation becomes more and more direct, and the gap between the highest state populated with appreciable probability and the energetic cutoff increases [Fig. 42(c)]. Rotational state distributions like the ones in Fig. 42(a) and 42(c) have been also obtained for HO_2 [60] and HNO [39] and we surmise that they are characteristic for dissociations on PES's, which have no barrier.

Experimental distributions measured for the two Λ-doublet states of

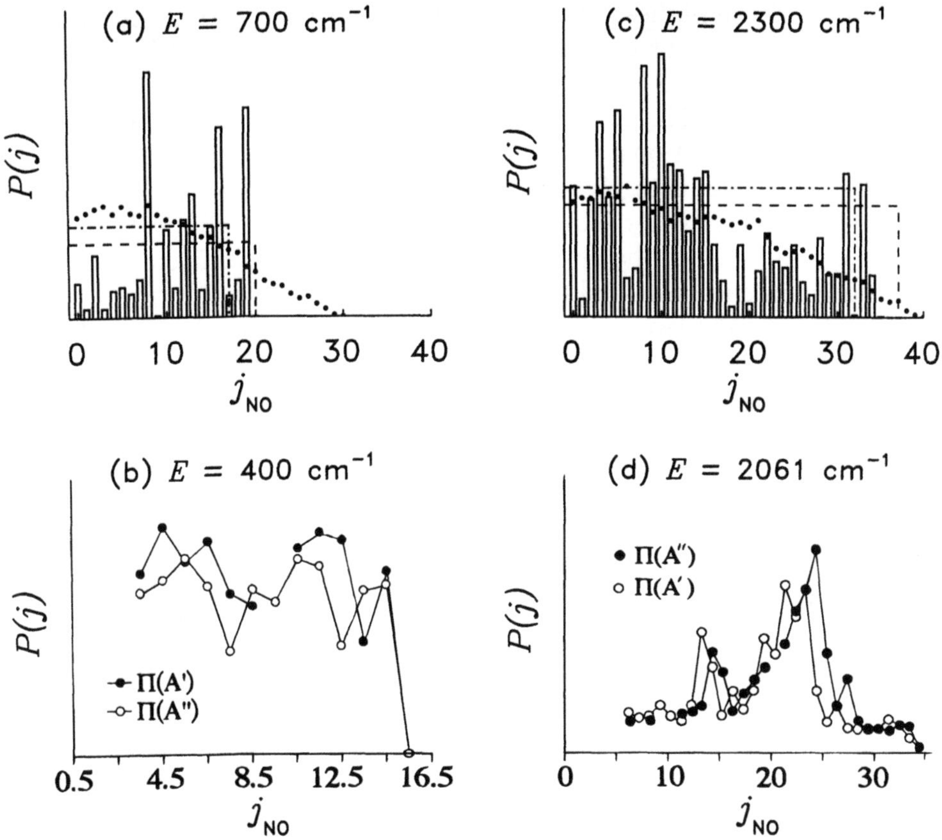

Figure 42: Theoretical [(a) and (c)] and experimental [(b) and (d)] rotational state distributions of NO in the dissociation of NO_2 at different excess energies. Upper panels: histograms – quantum mechanical calculations; full dots – classical trajectory calculations; dashed lines – PST; dashed-dotted lines – SACM. Lower panels: full and open symbols designate measured [133] distributions in the two Λ-doublet states of $NO(^2\Pi_{1/2})$.

$NO(^2\Pi_{1/2})$ are also presented in Fig. 42 (lower panels) for two excess energies close to those considered in the quantum mechanical studies. The calculated and experimental PSD's have the same *qualitative* appearance: erratic fluctuations up to j_{max} at low energies and slightly more structured distribtions which are 'colder' than j_{max} at higher energies. Of course, no quantitative agreement is expected at the present level of the quantum chemical and quantum mechanical calculations: The resulting distributions are far too sensitive to the fine details of the resonance wave functions — and, hence, of the underlying dynamics — to be correctly reproduced. Only the gross behaviour is of interest here.

The classical calculations reproduce — on the average — the quantum distributions quite well. At very low energies the classical distributions extend beyond the highest allowed state, because the zero-point energy of the NO fragment is not conserved and is partly transferred into rotation. With increasing E, however, this shortcoming becomes less severe. For the dissociation of HO_2 it has been shown that starting with randomly selected initial conditions at the TS and then following the trajectories outward to the products gives almost identical distributions [60] and the same is expected for NO_2. This means that, first, the statistical hypothesis is realistic and, second, the interaction in the exit channel is correctly taken into account in such classical calculations.

The statistical theories reproduce the quantum mechanical distributions rather accurately. In the phase space theory (PST) of Pechukas et al. [309,319], the coupling between the atom-diatom translational degree of freedom and the rotation of the product molecule is assumed to be strong, which implies that all asymptotically accessible states are equally populated. Thus, PST always gives the 'hottest' possible distribution. In the SACM, the coupling is taken to be zero from the TS to the infinitely separated products, and only those rotor states which are accessible at the TS are assumed to be populated. For this reason, SACM distributions are 'colder' than the PST ones. At low energies, both predictions nearly coincide [Fig. 42(a)], because the adiabatic curves are barrierless and the number of open states at the TS does not differ much from the number of open states at inifinite separation. At higher energies [Fig. 42(c)], the gap between the two statistical distributions widens, while the quantum results lie between the two predictions: Exit channel interactions are not as strong as predicted by PST, but are not vanishing either. Statistical models which include to some extent exit-channel coupling have been proposed by Bonnet, Rayez and coworkers [320–322].

Another molecule, for which rotational state distributions of the fragments have been intensively investigated, is hydrogen peroxide, H_2O_2, dissociating into OH + OH. It was studied by the direct overtone excitation by Crim and coworkers [143,144] and by infrared-optical double resonance by Luo and Rizzo [146]. In the latter study — similar to the work on HOCl reviewd earlier in this chapter — single rovibrational states of highly excited H_2O_2 are dissociated. As typical for a dissociation on a PES without barrier, the rotational distribution becomes gradually 'hotter' with increasing excess energy, that is, the highest rotational state populated successively increases and the maximum of the distribution shifts to higher and higher j values. The experimental data are in excellent agreement with the predictions of PST. Decisive for the success of the statistical model was the assumption that only the total angular momentum of OH is conserved, whereas the electronic angular momentum is not conserved during dissociation [146].

8. CLASSICAL CALCULATIONS

The kind of quantum mechanical studies described in the previous sections can be performed only for tri- or at most tetratomics. For larger systems, a classical mechanics treatment is the only alternative beyond statistical approaches. The general aspects of classical dynamics studies and several examples will be discussed in this section.

Initiated by the pioneering work of Bunker [323, 324] classical trajectory simulations have been extensively used to study the decomposition of energized molecules. In a unimolecular classical trajectory study, the motions of atoms for an ensemble of molecules are simulated by solving their classical equations of motion, usually in the form of Hamilton's equations, i.e.,

$$\frac{\partial H}{\partial q_i} = -\frac{\partial p_i}{\partial t} \quad \text{and} \quad \frac{\partial H}{\partial p_i} = \frac{\partial q_i}{\partial t}, \tag{60}$$

where the sum of the molecule's kinetic, $T(\mathbf{p}, \mathbf{q})$, and potential, $V(\mathbf{q})$, energies is the system's Hamiltonian

$$H = T(\mathbf{p}, \mathbf{q}) + V(\mathbf{q}) \,. \tag{61}$$

The kinetic energy can depend on both the momenta $\mathbf{p}$ and coordinates $\mathbf{q}$.

As discussed in the previous sections, to have the most complete comparison with experiment it is important to use an accurate potential energy function $V(\mathbf{q})$, usually obtained from high-level electronic structure theory calculations. In the past, this potential has always been represented by an analytical function. However, with the extraordinary increase in computer speed and enhancement in computer algorithms, it is now possible to calculate trajectories with the derivatives of the potential, $\partial V/\partial q_i$, in Eq. (60), obtained directly from an electronic structure theory, without the need for an analytical functional fit. Such 'on-the-fly' calculations are called *direct dynamics* classical trajectory simulations [325]. They are particularly important for molecules with many degrees of freedom, for which the construction of analytical PES's is impossible.

A power of classical trajectories is that they may be used in a 'pure simulation mode' [326] to investigate how changes in PES properties affect a molecule's unimolecular dynamics. Such a study provides fundamental insight into the relationship between the nature of intramolecular and unimolecular dynamics and the potential $V(\mathbf{q})$. The earliest classical trajectory studies [323, 324] were simulations of this type and they remain a powerful approach for studying unimolecular dynamics. It has provided detailed insight into the manner in which a molecule's intramolecular motion changes from regular to irregular [327] as more mode-mode coupling is included in $V(\mathbf{q})$ [328].

Once a potential energy function is chosen or determined for a molecule, there are three major components to a trajectory study: the selection of initial conditions for the excited molecule, the numerical integration of the classical equations of motion, and the analysis of the trajectories and their final conditions. The last item may include the time at which the trajectory decomposed to products, the nature of the trajectory's intramolecular motion, i.e., regular or irregular, and the vibrational, rotational and translational energies of the reaction products.

Unimolecular reactants are energized by a variety of experimental techniques including collisional and chemical activation, internal conversion and intersystem crossing transitions between electronic states, and different photo-activation techniques, which include excitation of isolated resonance states for reactants with a low density of states (see also Sect. 3). Trajectory simulations usually begin with the preparation of an ensemble of trajectories, whose initial coordinates and momenta resemble — as close as possible — those realized in a particular experiment [20, 329].

By choosing the initial conditions for an ensemble of trajectories to represent a quantum mechanical state, trajectories may be used to investigate state-specific dynamics and some of the early studies actually probed the possibility of state specificity in unimolecular decay [330]. However, an initial condition studied by many classical trajectory simulations, but not realized in any experiment is that of a micro-canonical ensemble [331] which assumes each state of the energized reactant is populated statistically with an equal probability. The classical dynamics of this ensemble is of fundamental interest, because RRKM unimolecular rate theory assumes this ensemble is maintained for the reactant [6, 332] as it decomposes. As a result, RRKM theory 'rules-out' the possibility of state-specific unimolecular decomposition. The relationship between the classical dynamics of a micro-canonical ensemble and RRKM theory is the first topic considered here.

8.1. Classical dynamics of a micro-canonical ensemble: intrinsic RRKM and non-RRKM behavior

The unimolecular rate constant $k(E)$, for a micro-canonical ensemble of reactant states, is identical with the RRKM rate constant. If $N(0)$ is the number of reactant molecules excited at $t = 0$ in accord with a microcanonical ensemble, the RRKM rate constant is then defined by

$$-\frac{dN(t)}{dt} = k(E)N(t) \tag{62}$$

evaluated at $t = 0$. The rate constant is expressed as the flux across the transition state separating reactants and products [333–336], i.e.,

$$k(E) = \frac{\iint d\mathbf{p}\, d\mathbf{q}\, \delta(H - E)\, \delta(q_1 - q_c)\, \dot{q}_1}{\iint d\mathbf{p}\, d\mathbf{q}\, \delta(H - E)}, \tag{63}$$

where q_1 is the reaction coordinate, q_c the value of the reaction coordinate at the TS and $\dot{q}_1$ the positive velocity in the reaction coordinate. By considering the relationship between the delta-function and the Heaviside function and integrating over q_1 and p_1 in Eq. (63), it is straightforward to show that Eq. (63) is equivalent to Eq. (42), i.e.,

$$k(E) = \frac{N_{\mathrm{TS}}(E)}{h\rho(E)}, \tag{64}$$

where $N_{\mathrm{TS}}(E)$ is the sum of states at the TS and $\rho(E)$ is the reactant's density of states. Equation (64) is the standard representation of the RRKM unimolecular rate constant and the quantum version is formulated by replacing the classical $N_{\mathrm{TS}}(E)$ and $\rho(E)$ by their quantum counterparts.

The fundamental assumption of RRKM theory is that the classical motion of the reactant is sufficiently chaotic so that a micro-canonical ensemble of states is maintained as the reactant decomposes [6, 324]. This assumption is often referred to as one of a rapid intramolecular vibrational energy redistribution (IVR) [12]. By making this assumption, at any time $k(E)$ is given by Eq. (62). As a result of the fixed time-independent rate constant $k(E)$, $N(t)$ decays exponentially, i.e.,

$$N(t) = N(0)\, e^{-k(E)t}. \tag{65}$$

A RRKM unimolecular system obeys the ergodic principle of statistical mechanics [337]. A quantity of more utility than $N(t)$, for analyzing the classical dynamics of a micro-canonical ensemble, is the *lifetime distribution* $P_{\mathrm{cl}}(t)$, which is defined by

$$P_{\mathrm{cl}}(t) = -\frac{1}{N(0)}\frac{dN(t)}{dt}. \tag{66}$$

According to RRKM theory, $P_{\mathrm{cl}}(t)$ is given by

$$P_{\mathrm{cl}}(t) = k(E)\, e^{-k(E)t}. \tag{67}$$

Both the intercept and slope of a plot of $\ln P_{\mathrm{cl}}(t)$ versus t give the RRKM rate.

The situation, where the actual classical $P_{\mathrm{cl}}(t)$ conforms to Eq. (67), is called *intrinsic* RRKM *behavior* [332]. *Intrinsic non*-RRKM *behavior* occurs when the energized molecule has a non-exponential $P_{\mathrm{cl}}(t)$, even if its internal vibrational states were prepared randomly. This situation arises when transitions between individual molecular rovibrational states are slower than transitions leading to products. As a result, the vibrational states do not have equal dissociation probabilities. In terms of classical phase space dynamics, slow transitions between the states occur when the reactant phase space consists of several regions, which are weakly coupled on the timescale of the unimolecular reaction, and when there is at least one bottleneck in the phase space other than the one defining the TS. Intrinsic non-RRKM molecules, excited with an initial micro-canonical ensemble of states, decay non-exponentially with a time-dependent unimolecular rate constant or exponentially with a rate larger than that of the RRKM theory. In the case of non-exponential decay, one distinguishes an initial fast decay (usually faster than the RRKM rate) and a slower decay component, which sets in at later times. For micro-canonical ensembles, the average of the fast and the slow decay rates equals the RRKM rate constant [6]. Exponential non-RRKM decay occurs when part of the phase space is totally decoupled from the reaction coordinate. Very often, the decay in this case is faster than k_{RRKM}, because the reactive part of the phase space is effectively smaller than the complete phase space [338].

Initiated by the work of Bunker [323,324], extensive trajectory simulations have been performed to determine whether molecular Hamiltonians exhibit intrinsic RRKM or non-RRKM behavior. Both types have been observed and in Fig. 43 we depict two examples, i.e., classical lifetime distributions for NO_2 [271] and O_3. While $P_{\mathrm{cl}}(t)$ for NO_2 is well described by a single-exponential function — in contrast to the experimental and quantum mechanical decay curves in Fig. 31 —, the distribution for ozone shows clear deviations from an exponential decay. The classical dynamics of NO_2 is chaotic, whereas for O_3 the phase space is not completely mixed. This is in accord with the observation that the quantum mechanical wave

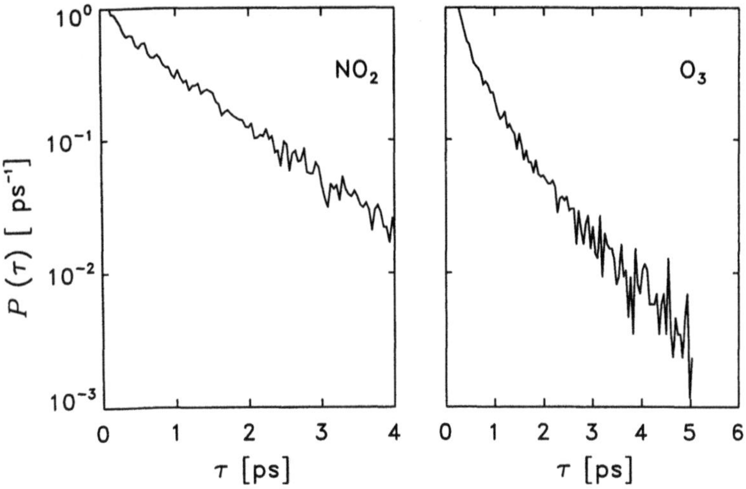

Figure 43: Classical lifetime distribution functions $P_{cl}(\tau)$ for NO_2 [271] and O_3. τ is the time at which the transition state has been reached. The energies above the classical thresholds are 550 cm^{-1} for NO_2 and 1 430 cm^{-1} for O_3.

functions of NO_2 are not assignable and that a significant number of resonance states for O_3 have a clear assignment, even at high energies above threshold [42,45]; the motion of ozone on its ground-state PES is not completely statistical. Not surprisingly, the molecules discussed in Sect. 5 such as HCO [51], which have mode-specific unimolecular dynamics, are intrinsically non-RRKM. Their $P_{cl}(t)$ do not follow a single exponential and may be represented as sums of several exponentials according to

$$P_{cl}(t) = \sum_i c_i \, e^{-k_i t} , \tag{68}$$

where, by definition, the sum of the c_i equals the RRKM rate constant, since the initial decay is for a micro-canonical ensemble.

Examples for reactions with intrinsic RRKM dynamics include $Al_n \rightarrow Al_{n-1} + Al$ [339], $HO_2 \rightarrow H + O_2$ [60], $C_2H_5 \rightarrow H + C_2H_4$ [338], $CH_4 \rightarrow H + CH_3$ [340]. The following (dissociation, isomerization and bimolecular) reactions have been shown to a have an intrinsically non-RRKM behavior: $CH_3NC \rightarrow CH_3CN$ [332], $HCO \rightarrow H + CO$ [51], $Cl^- + CH_3Cl \rightarrow ClCH_3 + Cl^-$ [341], $HC_2 \rightarrow H + C_2$ [328], $(CH_3)_2NNO_2 \rightarrow (CH_3)_2N + NO_2$ [342], the isomerization of tri-methylene [343, 344], $Si_2H_6 \rightarrow 2SiH_3$ [345] and the dissociation of hexahydro-1,3,5-trinitro-1,3,5-triazine (RDX) [346]. It should be noted that the last two reactions become intrinsically RRKM at low energies. In addition, Wales and coworkers [347] have studied the unimolecular dissociation of numerous clusters and macromolecules. In their studies, they have not clearly identified any example in which the decomposition from a single minimum of the cluster or macromolecule is non-exponential and, thus, intrinsically non-RRKM [348].

It is not immediately obvious, by simply 'looking at' a molecule's Hamiltonian and/or its PES, whether the unimolecular dynamics will be intrinsic RRKM or not and computer simulations as outlined here are required. Intrinsic non-RRKM dynamics is indicative of mode-specific decomposition, since different regions of phase space are not strongly coupled and a micro-canonical ensemble is not maintained during the fragmentation. The phase space structures, which give rise to intrinsic RRKM or non-RRKM behavior, are discussed in the next section.

8.2. Phase-space structures

As discussed above, the assumption of RRKM theory is that the internal motion of the excited molecules is irregular on the timescale of the unimolecular reaction, so that a micro-canonical ensemble is maintained within the phase space of the molecule. Thus, it is of interest to consider the properties of molecular Hamiltonians which give rise to ergodic and non-ergodic dynamics. Because the structure of the classical phase space is so important for understanding the bound-state spectrum of a molecule as well as its dissociation dynamics, it has been thoroughly investigated and many general conclusions have been drawn from these studies, although they dealt mainly with model rather than accurate calculated PES's [114, 349, 350]. In the following we will briefly review some of the conclusions.

8.2.1. Quasi-periodic and chaotic motions

For a very low energy, leading to infinitesimal displacements from the potential energy minimum, the molecule's energy is given by the separable normal-mode (nm) Hamiltonian

$$E = H_{\text{nm}} = \sum_i (P_i^2 + \lambda_i Q_i^2)/2 \equiv \sum_i E_i\,, \tag{69}$$

for which the energy is a sum of the energies in the individual normal modes (Q_i, P_i). The classical motion is regular [327], giving rise to quasi-periodic trajectories with no energy transfer between the modes. For quasi-periodic motion the trajectory moves in a restricted region of phase space which, according to a theorem in topology, has the shape of a n-dimensional torus [327].

As the energy is increased, a coupling term H' between the normal modes becomes important for the molecular Hamiltonian, i.e.,

$$H = H_{\rm nm} + H' \,, \tag{70}$$

but $H_{\rm nm}$ remains still a good zero-order representation of the full Hamiltonian. The additional coupling term gives rise to an additional type of classical motion called chaotic (irregular) [327]. For chaotic motion, energy flows freely between the molecule's modes and the molecule does not exhibit regular vibrational motion. At intermediate energies, chaotic trajectories coexist with quasi-periodic ones. A continued increase in energy leads to a gradual destruction of the remaining tori to which quasi-periodic trajectories are confined. If all the tori are destroyed, there are no quasi-periodic trajectories, all regions of phase space become accessible to a trajectory, and the ergodic hypothesis becomes valid.

A Poincaré surface of section may be used to identify the chaotic and quasi-periodic regions of phase space for a two-dimensional Hamiltonian. An ensemble of trajectories, chosen to randomly sample the phase space, are calculated and for each trajectory a point is plotted in the (q_1, p_1)-plane every time $q_2 = 0$ for $p_2 > 0$. A quasi-periodic trajectory lies on an invariant curve, while the points are scattered for a chaotic trajectory with no pattern. Figure 44 shows an example for a two-dimensional model for HOCl; the HO bond distance is frozen in these calculations [351]. It clearly illustrates how the phase space becomes gradually more chaotic as the energy increases.

Poincaré surfaces of section are difficult to construct and to interpret for Hamiltonians with more than two degrees of freedom and other procedures must be used to identify a trajectory as quasi-periodic or chaotic. One approach is to calculate the power spectrum of a trajectory given by the Fourier analysis [352] according to

$$I(\omega) = \frac{1}{2\pi} \lim_{T\to\infty} \frac{1}{2T} \left| \int_0^{2T} dt \sum_i q_i(t)\, e^{-i\omega t} \right|^2 . \tag{71}$$

For quasi-periodic trajectories, like those for the normal-mode Hamiltonian in Eq. (69), $I(\omega)$ consists of a series of lines at the frequencies for the normal modes of vibration. In contrast, a Fourier analysis of a chaotic trajectory results in a multitude of peaks, without identifiable frequencies for particular modes. An inconvenience in this approach is that for a large molecule with many modes, a trajectory may have to be integrated for a long time T to resolve the individual lines in a power spectrum for a quasi-periodic trajectory. Moreover, in the presence of a resonance between different modes, the interpretation of the power spectrum may become misleading.

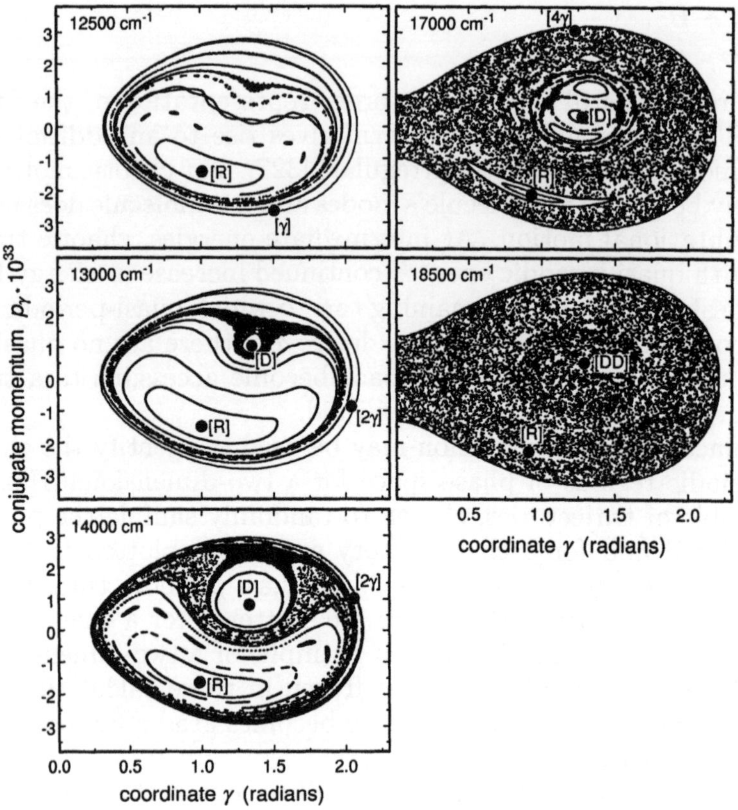

Figure 44: Poincaré surfaces of section for a two-dimensional model of HOCl; the HO bond distance is frozen in the classical calculations. γ is the Jacobi angle and p_γ is the corresponding momentum. Different symbols in square brackets denote different types of periodic orbits in the molecule. Reproduced, with permission of the American Chemical Society, from Ref. 258.

The nature of the intramolecular motion may also be identified by studying the way the separation of two trajectories evolves with time [353]. If the motion is regular (quasi-periodic) the separation is linear with time, but exponential if the motion is irregular (chaotic). If the separation is exponential, the rate of the separation — called the Lyapunov characteristic exponent — provides qualitative information concerning the IVR rate for the chaotic trajectories. This type of analysis has been reported, for example, for NO_2 [271] and the $Cl^- \cdots CH_3Br$ complex [354].

The dynamics for a particular type of vibration, e.g. O–H stretch in H_2O, undergoes a transition from a quasi-periodic to chaotic motion as the energy is increased. At energies where this transition first begins to occur, the nature of the motion may be quite complex, appearing quasi-periodic for short times, but then becoming chaotic-like. The term 'vague torus' has been coined for this situation [355]. In addition, at even higher energies, when the motion appears to be chaotic, with careful analysis it has been possible to characterize the remnants of quasi-periodic trajectories in the classical phase space [356].

Especially important for unimolecular dissociation studies is that quasi-periodic trajectories can exist above the dissociation threshold [328]. Because of the quasi-periodic motion, the trajectory cannot reach the TS separating the bound region from the product channel and remains trapped in the reactant region of phase space. This type of behavior was observed in a trajectory study based on a three-dimensional Hamiltonian for the $HC_2 \rightarrow H + C_2$ model system [328]. Examples, investigated more recently, are the dissociation of HOCl [351] and O_3 [45]. In the latter case, periodic orbits for the bending and the symmetric stretching modes have been found for energies high above the dissociation threshold. In the quantum mechanical world, periodic or quasi-periodic trajectories correspond to assignable wave functions. Illustrative examples, which have been discussed in this chapter, are HCO and HOCl.

8.2.2. Transition from quasi-periodic to chaotic motion

In the above it is described that a molecular Hamiltonian undergoes a transition from quasi-periodic to mixed quasi-periodic and chaotic and then ergodic behavior as the energy is increased. It is important to understand this transition. Numerical simulations have shown that the presence of the perturbation H' in Eq. (70) is not sufficient for ergodic dynamics. It was discovered that the vibrational degrees of freedom will not freely exchange energy as long as there are no resonance conditions for energy transfer [327], specified by $n_1\omega_1 - n_2\omega_2 \approx 0$ for a two-dimensional system. These frequency matchings are known as 'internal' or 'anharmonic resonances' and the lack of these resonances precludes appreciable energy sharing in the weak coupling limit. The reason for this result is explained by the Kolmogorov-Arnold-Moser (KAM) theorem [114], which states that if there are no resonances among a set of oscillators, the addition of a perturbation, that is sufficiently small compared to the total energy, does not make the system ergodic.

The theorem can be exemplified by considering the effect of the perturbation on the phase-space tori. Tori of the unperturbed Hamiltonian with incommensurate frequencies continue to exist, being only slightly distorted by the perturbation. On the other hand, the tori with commensurate frequencies, or with incommensurate frequencies whose ratio is well approximated by a ratio of small integers are grossly deformed. The KAM theorem shows that, when H' is sufficiently small, the majority of initial conditions for the full Hamiltonian lie on the preserved tori and non-ergodic dynamics results. This justifies the view that the perturbation H' largely serves only to slightly change the frequencies. However, not all tori survive. Those, which are destroyed, form regions in the phase space, which give rise to chaotic motion. These regions are intermingled with the preserved tori and for systems with three and more degrees of freedom they build an Arnold web [357, 358] of internal chaotic resonances. It is the existence of this relatively small set of destroyed tori which ultimately lead to ergodic behavior as the energy increases [358, 359].

These considerations are, strictly speaking, only applicable for the classical phase space. Of course, in quantum mechanics one finds the equivalent picture: assignable states at low energies, assignable and non-assignable states coexisting at intermediate energies, and mainly unassignable states at very high energies. NO_2 is an illustrative example. However, the energy regime where the transition to fully irregular classical motion occurs may be different from the energy region where all quantum states become unassignable. For example, in a two-dimensional model for HOCl the classical phase space at energies near the dissociation threshold is fully chaotic, whereas almost all quantum wave functions are still assignable at these energies [351].

8.2.3. Models for non-ergodic dynamics

The simplest approach [338] to describe a non-ergodic unimolecular system is to assume that the reactant's phase space only consists of quasi-periodic and chaotic trajectories, whose numbers are N_{qp} and N_{ch}. If a micro-canonical ensemble is prepared at $t = 0$ and if it is assumed that a restricted micro-canonical ensemble is maintained within the chaotic region, while no trajectory dissociates from the quasi-periodic region, the number of reactant molecules versus time is

$$N(t) = N_{\text{qp}}(0) + N_{\text{ch}}(0)\, e^{-k_{\text{ch}} t}\,, \tag{72}$$

where k_{ch} is the rate constant for the chaotic region and the rate constant is zero for the quasi-periodic region. Since there is an initial micro-canonical ensemble in the reactant's phase space,

$$-\frac{dN(t)}{dt} = k_{\text{RRKM}}\, N(0) = k_{\text{ch}}\, N_{\text{ch}}(0) \tag{73}$$

evaluated at $t = 0$. Thus, $k_{\text{ch}} = k_{\text{RRKM}}/F_{\text{ch}}$, where F_{ch} is the chaotic fraction of phase space, i.e., $N_{\rm ch}(0)/N(0)$. The unimolecular lifetime distribution, Eq. (66), is

$$P_{\rm cl}(t) = k_{\rm RRKM}\, e^{-k_{\rm ch}t}\,. \tag{74}$$

This model is clearly incomplete, since it does not account for vague tori [355] and the complex Arnold web [357, 358] structure of a multidimensional phase space with both chaotic and quasi-periodic trajectories. However, Eq. (74) does properly describe that, with non-ergodic dynamics, the lifetime distribution will have an initial component that decays faster than the RRKM prediction as found in the simulations by Bunker [323,324] and the more recent study of HCO dissociation [51]. Additionally, there will be a component to the classical rate, which is slower than $k_{\rm RRKM}$, for example, in the dissociations of NO_2 and O_3; this component cannot be described by an expression as simple as the one in Eq. (74).

For two-dimensional classical systems it is possible to calculate the non-ergodic rate constant from the phase-space structure [360, 361]. These systems have invariant tori known as KAM-surfaces, which serve as absolute boundaries in phase space through which no phase-space points pass. At energies above the unimolecular threshold, the strength of perturbation parameters increases, and KAM-surfaces may be destroyed and no longer serve as absolute boundaries. However, remnants of them, termed cantori because they form Cantor sets, may still survive as potential barriers and form phase-space bottlenecks. Applications of such concepts, as a modification to RRKM theory, have been successful in treating two-mode systems, e.g., $HeI_2 \rightarrow He + I_2$ [361]. Generalization of this approach to higher dimensions is uncertain, since KAM-surfaces will no longer serve as boundaries.

A classical diffusion theory model has been proposed to calculate the rate of IVR between the reaction coordinate and the remaining 'bath' modes of the molecule [345]. Following work by Bunker [324], the unimolecular dynamics will be non-ergodic (intrinsically non-RRKM) if $k_{\rm RRKM} \gg k_{\rm IVR}$. For such a situation, the unimolecular decomposition will be exponential and occur with a rate constant equal to $k_{\rm IVR}$. The rate of IVR is modeled by assuming a random force between the bath modes and the reaction coordinate. The model was used to successfully analyze the intrinsic non-RRKM dynamics for $Si_2H_6 \rightarrow 2\,SiH_3$ dissociation [345].

A complete model for the non-ergodic classical dynamics of a polyatomic molecule will need to represent the complete Arnold web structure of the phase space. There may be multiple bottlenecks for IVR and vague tori may exist in the vicinity of invariant tori. These complex phase space structures, leading to non-ergodic dynamics, are the origins of the non-exponential decomposition found for a micro-canonical ensemble of HCO [51] or O_3 (see Fig. 43) molecules. It is not difficult to envision that such models will not be simple.

8.3. Simulations of experimental molecular systems with non-random excitation

The situation considered above, in which the unimolecular reactant is initially excited with a micro-canonical distribution of states, is an idealized one and does not occur in any experimental excitation process, including collisional activation. The actual lifetime distribution for the unimolecular reactant will depend in part on the manner in which the energy needed for the reaction is supplied. In many experiments, such as photo- and chemical activation, the initial distribution of excited molecular rovibration states is distinctly non-random. Regardless of the pattern of the initial energizing, the RRKM model of rapid IVR requires the distribution of states to become micro-canonical in a negligibly short time. Though the excitation is non-random, the lifetime distribution may be similar to that of RRKM theory. In other cases, the probability of a short lifetime with respect to reaction may be enhanced or reduced, depending on the properties of the initial wave packet and the location of the excitation within the molecule. These are examples of *apparent* non-RRKM behavior [332] arising from the initial non-random excitation. If there are strong internal couplings between the vibrational modes, $P_{\mathrm{cl}}(t)$ will become that of RRKM theory, Eq. (67), after rapid IVR. In contrast, if the molecule decomposes via mode-specific resonance states, the resulting $P_{\mathrm{cl}}(t)$ is expected to strongly depend on the region of phase space initially excited. In the following we will review two examples.

Trajectory simulations have been used to study the unimolecular dynamics of ethyl radicals excited non-randomly by the $H + C_2H_4 \rightarrow C_2H_5$ chemical activation mechanism [338]. Initial conditions were chosen for an ensemble of trajectories with ethylene containing no rotational energy and in its zero-point vibrational energy level. The initial relative translational energy is 30 kcal/mol, C_2H_4 has a random orientation with respect to the colliding H atom and the collision impact parameter is chosen randomly. The intramolecular and unimolecular dynamics of the excited C_2H_5 radicals was followed for 1 ps.

The dynamics of the $H+C_2H_4$ association step deposits the 40 kcal/mol reaction exothermicity and the 30 kcal/mol relative translational energy non-randomly in the energized C_2H_5 radical, with the C–H bond, that is formed, preferentially excited as compared to the other C–H bonds. The expectation is that this non-random excitation may lead to an initial dissociation rate that is larger than that of RRKM theory. However, C_2H_5 is

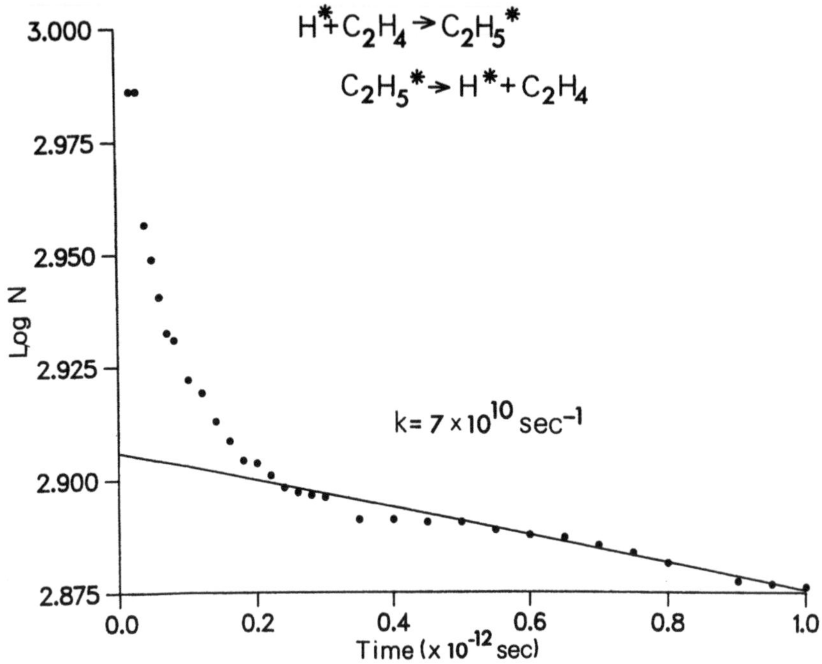

Figure 45: Classical lifetime distribution for C_2H_5 decomposition. Reproduced, with permission of the American Chemical Society, from Ref. 338.

an intrinsic RRKM molecule and rapid IVR will quickly lead to a dissociation rate constant in agreement with that of RRKM theory. As shown in Fig. 45, this is what is found from the trajectory simulations. There is an apparent non-RRKM component to the dissociation for the first 0.2 ps and then the rate constant (i.e., the slope of $\log N$ versus t) becomes 7×10^{10} s^{-1}, which agrees with the statistical dissociation rate constant for a micro-canonical ensemble of energized C_2H_5 radicals.

Both trajectory and experimental studies [362, 363] have shown that the unimolecular dynamics associated with S_N2 reactions of the type $X^- + CH_3Y \rightarrow XCH_3 + Y^-$ is state-specific and intrinsically non-RRKM. These reactions are characterized by double-well potentials (shown in Fig. 46 for $Cl^- + CH_3Cl$ and $Cl^- + CH_3Br$). The ion-dipole minima correspond to $X^- \cdots CH_3Y$ and $XCH_3 \cdots Y^-$, and the intra- and intermolecular modes in the wells are weakly coupled. For the $X^- \cdots CH_3Y$ minimum, the CH_3Y vibrational modes are intramolecular, while the three stretching and bending modes associated with the relative motion of X^- with respect to CH_3Y are intermolecular. The non-RRKM dynamics for the ion-dipole complexes give rise to a number of important attributes for the $X^- + CH_3Y$ reactions: (1) a lifetime for the $X^- \cdots CH_3Y$ complex much longer than the

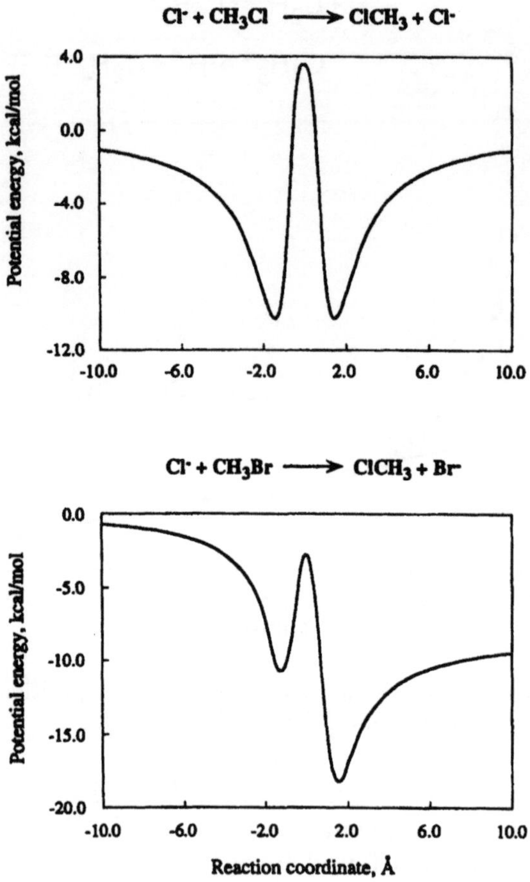

Figure 46: Representative reaction path potentials for $Cl^- + CH_3Cl \rightarrow ClCH_3 + Cl^-$ and $Cl^- + CH_3Br \rightarrow ClCH_3 + Br^-$ nucleophilic substitution. Reproduced, with permission of the American Association for the Advancement of Science, from Ref. 362.

prediction of RRKM theory [364]; (2) rate constants for the $X^- + CH_3Y \rightarrow XCH_3 + Y^-$ S_N2 reaction, whose dependence on reagent translational, vibrational, and rotational energies disagree with RRKM theory [365]; (3) a direct S_N2 reaction mechanism, without trapping in either ion-dipole potential energy minimum [362]; (4) trajectories trapped in the $[X \cdots CH_3Y]^-$ central barrier region of the PES, with extensive barrier re-crossings [366], i.e., a violation of RRKM theory; and (5) $XCH_3 + Y^-$ product energies in disagreement with statistical energy partitioning [367].

Several illustrative features of the dynamics of the $Cl^- + CH_3Cl$ reactive system are presented here. Simulations have been performed for collisions between Cl^- and CH_3Cl and the lifetimes of the $Cl^- \cdots CH_3Cl$ complexes, formed in these collisions, have been recorded [341]. Lifetime

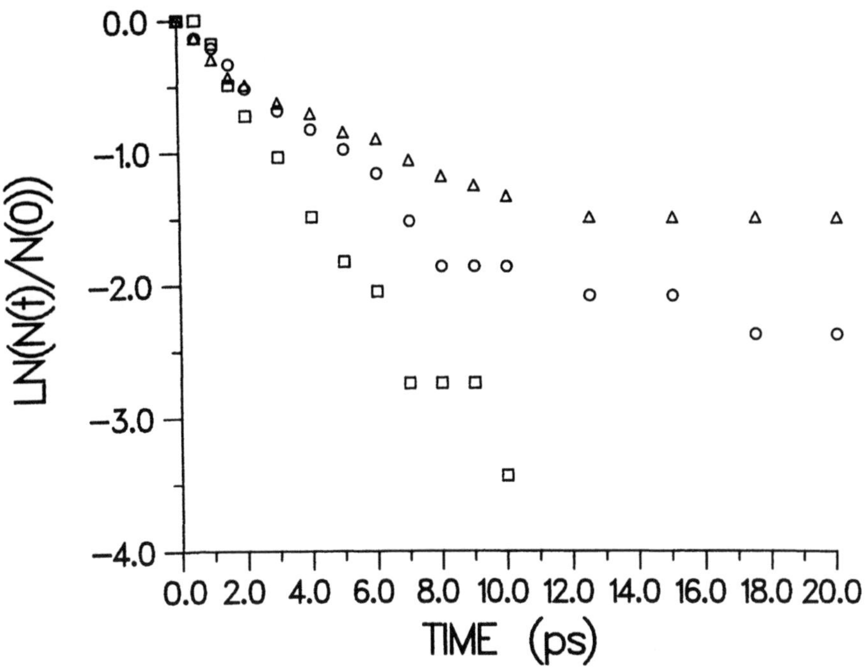

Figure 47: Number of $Cl^- \cdots CH_3Cl$ complexes, formed by $Cl^- + CH_3Cl$ collisions, with a relative translational energy of 2.0 kcal/mol versus time. The CH_3Cl reactant has no energy except n quanta in the C–Cl stretching mode: $n = 6$ (triangles); $n = 3$ (circles); $n = -0.5$ (squares). Reproduced, with permission of the American Physical Society, from Ref. 341

distributions are shown in Fig. 47 for the complexes formed from collisions, at a relative translational energy of 2.0 kcal/mol, of Cl^- with CH_3Cl molecules containing no vibration/rotation energy, except excitation of the C–Cl stretch normal mode. The ensemble of trajectories is averaged over the collision impact parameter and random orientations of the CH_3Cl molecules. The lifetimes in Fig. 47 illustrate extensive non-RRKM dynamics. Remarkably, the lifetimes of the complexes become longer when the energy of the CH_3Cl molecule is increased by adding quanta to the C–Cl stretch mode. Thus, adding energy to the complex increases its lifetime, a result contradictory to the RRKM prediction. This result may be qualitatively understood by considering IVR for the $Cl^- \cdots CH_3Cl$ complex and its possible phase-space structure. Analysis of the trajectories show that, when the complexes are initially formed, only their intermolecular modes are excited (i.e., the $Cl^- \cdots C$ stretch and the two $Cl^- \cdots CH_3Cl$ bending modes. Adding energy to the C–Cl stretch of CH_3Cl, apparently makes the intramolecular motion more irregular, enhancing energy transfer from the intermolecular modes to the CH_3Cl intramoleuclar modes, which would increase the lifetime of the complexes. It is noteworthy that, when calculations for different energies are averaged in order to simulate collisions at 300 K, the lifetime of $Cl^- \cdots CH_3Cl$, determined from the trajectory lifetime distribution, agrees with the experimental result [364].

Non-RRKM dynamics for the $Cl^- + CH_3Cl$ reactive system are also illustrated by trajectories initiated at the $[Cl \cdots CH_3 \cdots Cl]^-$ central barrier, which were first studied with an analytic PES fit to HF/6-31G* ab initio calculations [367] and more recently investigated by direct dynamics at the MP2/6-31G* level of theory [366]. Both studies give the same results. The height of the central barrier, with respect to the reactant and product asymptotic limits, is sufficiently large so that transition state theory predicts crossing the barrier to be rate determining, with negligible barrier re-crossing. Furthermore, RRKM theory assumes that when the reactive system passes the central barrier, it forms a long-lived $ClCH_3 \cdots Cl^-$ comples in the exit channel, which has an RRKM lifetime and dissociates to the $ClCH_3 + Cl^-$ products. A mirror image of these dynamics is predicted on the reactant side of the barrier.

The actual dynamics are decidedly different than that of the transition state and RRKM theoretical models. An example trajectory from the latter direct dynamics study is shown in Fig. 48. The intramolecular motion is regular, not irregular as predicted by RRKM theory. Given the regular motion, more barrier crossings are expected if the trajectory were followed for more than 3 ps in both the forward and backward directions off the barrier. None of the trajectories reached either the product or reactant asymptotic limit within 3 ps, even though RRKM theory predicts 50% reaction in this time interval. The ratio of the number of reactions to barrier crossings is less than 0.2, while transition state theory assumes this ratio is unity. A recent quantum dynamical study of the $Cl^- \cdots CH_3Cl$ complex, in which, however, only the two C–Cl stretching coordinates in a collinear arrangement are taken into account [78], also found regular dynamics, in agreement with the trajectory studies described here.

8.4. Direct dynamics simulations

As discussed at the beginning of this section, with recent advances in electronic structure theory algorithms and the significant increase in computer speed, it is now possible to calculate classical trajectories without the need for an analytic potential energy function. The potential energy and its gradient, required for calculating a trajectory, are obtained 'on-the-fly' from an electronic structure theory [325]. Two methods have been advanced for performing such direct dynamics. In the Car-Parrinello approach [368], the motions of the electrons are determined simultaneously

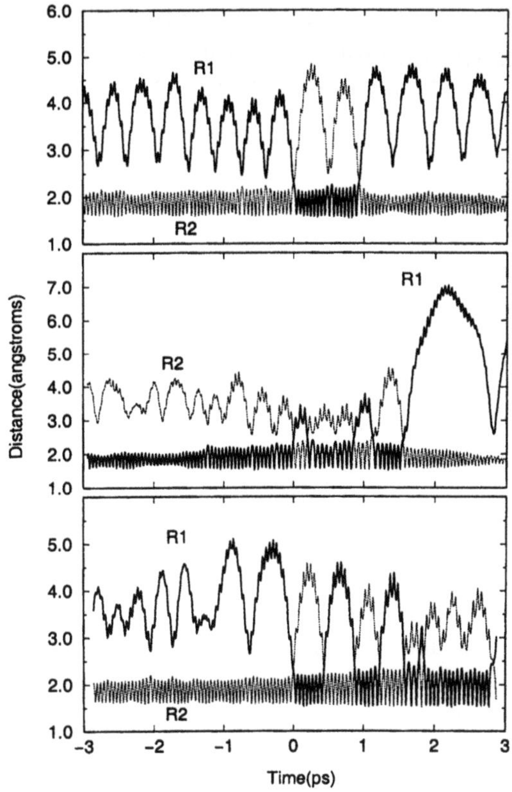

Figure 48: Representative trajectories for excitation at the $[Cl \cdots CH_3 \cdots Cl]^-$ central barrier. The trajectories are initiated at time equal zero and integrated for 3 ps in both forward and backward directions. The dark line is R_1 and the light line R_2, the two C–Cl bond lengths. Reproduced, with permission of the American Chemical Society, from Ref. 366.

as the nuclear classical equations of motion are integrated, to determine the change in the electronic wave function as the nuclei move. For the second method, Born-Oppenheimer direct dynamics, the electronic wave function is optimized at each numerical integration step of the classical trajectory [325]. This latter approach was used for the direct dynamics trajectory simulations discussed here.

A strength of direct dynamics is that it allows one to determine the classical dynamics for a particular level of electronic structure theory, without the need for an intermediate analytic PES fit. In addition, since it is difficult to develop analytic PES's for many-atom systems, direct dynamics greatly expands the range of unimolecular systems whose dynamics may be investigated by classical trajectories. However, a limitation of direct dynamics, is the large computer time needed to calculate the trajectories. For example, direct dynamics at the MP2/6-31G* level of theory was used to investigate the unimolecular and central barrier dynamics for the $Cl^- + CH_3Cl$ S_N2 reaction discussed in 8.3 [366]. Approximately one week of CPU time on one processor was required to integrate a single trajectory for 3 ps in both the forward and reverse directions off the reaction's central barrier. The CPU time would be even greater if a higher level of electronic structure theory was used. The substantial requirement of CPU time for a direct dynamics simulation may be addressed by either a large cluster of processors, each calculating one trajectory, or a very efficient distributed computing environment, which allows the calculation of an ensemble of trajectories at one time as well as distributing the calculation of a single trajectory over multiple processors. In addition to the above $Cl^- + CH_3Cl$ simulation, direct dynamics have been used to study other S_N2 reactions, including $F^- + CH_3Cl$ [369], $F^-(H_2O) + CH_3Cl$ [370], $Cl^- + CH_3Br$ [371], micro-solvated $Cl^- + CH_3Br$ [372], and $Cl^- + ClCH_2CN$ [373], and a range of non-statistical dynamics have been identified in these studies. To illustrate the broad applicability of direct dynamics to unimolecular systems, in the following, two additional applications of direct dynamics are discussed. One is a study of product energy partitioning in unimolecular dissociation and the other is unimolecular fragmentation in surface-induced dissociation (SID).

Direct dynamics is particularly applicable to unimolecular processes which are direct and require very little time. An example is the formation of unimolecular fragments as a reactive system moves from the dissociation barrier to products. If the nature of the trajectory's initial conditions at the barrier is known, direct dynamics may be used to calculate the vibrational, rotational and translational energies of the unimolecular products. RRKM theory is valid for many unimolecular reactions and for these cases a micro-canonical ensemble may be sampled at the barrier, with its rovibrational states having equal probabilities of being populated. This sampling procedure has been used, for example, in direct dynamics simulations to determine the product energy partitioning for $C_2H_4F \rightarrow C_2H_3F + H$ [374] and $H_2CO \rightarrow H_2 + CO$ [375] dissociations. Very good agreement is found between the simulations and experiment [312]. This is illustrated in Table 1, which gives the vibrational distributions of the H_2 and CO products for H_2CO dissociation. The direct dynamics calculations have been performed on the Hartree-Fock level using two different basis sets.

Product energy partitioning in the dissociation of ketene in the S_0 state, $CH_2CO \rightarrow CH_2(^1A_1) + CO$, has also been studied by direct dynamics [376]. Here the trajectories are initialized at the variational transition state separating reactants and products. The rotational energy distributions of the CH_2 and CO products are in good agreement with experiment and different

Table 1: Calculated (Ref. 375) and measured (Ref. 312) relative populations of the vibrational product states of CO and H_2 following the dissociation of H_2CO

	CO			H_2		
n	HF/3-216	HF/6-31G*	exp.	HF/3-216	HF/6-31G*	exp.
0	83.2	82.2	88	11.0	22.8	24.2
1	16.8	17.8	12	30.7	36.5	41.3
2				29.5	27.0	24.6
3				21.7	11.2	8.6
4				6.8	2.5	0.3
5				0.3	0.0	0.0

than the PST prediction. However, the simulations do not reproduce the correlations between the CH_2 and CO rotational states. Additional work needs to be done to explain this discrepancy.

Direct dynamics is applicable to large molecular systems, but a lower level of electronic structure may be required as well as a 'blend' of direct dynamics and analytic potential energy functions. This latter technique, often called quantum mechanical/molecular mechanical (QM/MM) direct dynamics [377], has been used to simulate SID unimolecular dynamics associated with protonated glycine ions, $NH_3CH_2COOH^+$ [(gly–H)$^+$], colliding with a hydrogenated diamond {111} surface [378]. The potential energy for the system is represented by

$$V = V_{\text{gly}} + V_{\text{surf}} + V_{\text{surf-gly}}\,, \tag{75}$$

where V_{gly}, the (gly–H)$^+$ intramolecular potential, is represented by the AM1 semiempirical electronic structure theory, V_{surf} is an analytic potential energy function for the diamond surface, and $V_{\text{surf-gly}}$ is an analytic function for the intermolecular potential between (gly–H)$^+$ and the surface. The QM direct dynamics component of the simulation is the (gly–H)$^+$ potential energy and gradient obtained from AM1 theory.

A very important finding from this study is a shattering mechanism, in which (gly–H)$^+$ fragments to products during the collision with the surface. If (gly–H)$^+$ is properly oriented, as it collides with the hard diamond surface, it is driven to a transition state structure and fragments. For (gly–H)$^+$ collisions with diamond {111} at an energy and angle of 70 eV and 45°, approximately 55% of the dissociations occur by shattering. An important (gly–H)$^+$ fragmentation channel is the formation of the $NH_2CH_2^+ + C(OH)_2$ products. The reaction is initiated when the NH_3 moiety of (gly–H)$^+$ collides with the surface, causing the transfer of a H-atom from NH_3 to the carbonyl oxygen during the collision.

8.5. Expected accuracy of classical calculations

Classical trajectory calculations are an indispensable tool in studies of molecular dynamics, for small and, even more so, for large systems. Therefore it is essential to investigate the accuracy one can expect from classical mechanics. In principal, this can only be done by comparison with exact quantum mechanical calculations, which, however, are limited to relatively small molecules. Numerous comparisons in the context of energy transfer and exchange reactions [379] in bimolecular collisions as well as photodissociation [20], have shown that classical calculations are well suited to reproduce the gross features of e.g. collision cross sections. More subtle effects like interferences and tunneling intrinsically cannot be correctly described by pure classical mechanics, unless 'phases' associated with each trajectory are also incorporated, for example in the semiclassical initial-value-representation [380]. For large systems it is more difficult to assess the accuracy of classical mechanics, because quantum mechanical calculations including all degrees of freedom are not manageable.

In this section we will briefly consider a few questions concerning the expected accuracy of trajectory calculations in the context of unimolecular dissociation processes. For a particular molecular system it depends on the size of the system (number of atoms), the energy regime, the nature of the classical motion (e.g. regular or chaotic) and, very importantly, the timescale. These issues will be addressed in the following discussions of both random and non-random excitation of molecules.

8.5.1. Decomposition of a micro-canonical ensemble — small molecules

Comparisons between state-specific quantum mechanical and classical calculations have been made for four systems, HO_2 [60], NO_2 [271], HNO [39], and HCO [51]. For the first three systems the quantum dynamics is statistical state-specific and the classical dynamics is in essence irregular above the dissociation threshold; HCO is an example of mode-specific quantum mechanical behavior and the classical phase space is certainly not completely chaotic.

If the classical dynamics is ergodic and intrinsically RRKM, one might expect that the classical rate constant approximates the average rate of the quantum mechanical state-specific rates. That is indeed the case for the dissociation of HO_2 (Fig. 12 of Ref. 60); the classical rate is only slightly smaller than the average quantum mechanical rate. The same holds also for NO_2; if the classical rate constants calculated in Ref. 271 were included in Fig. 30, it would agree well with the average quantum rate as well as the experimental rate constants, except for the threshold region (see below). Both for HO_2 and NO_2 the lifetime distribution function, $P_{cl}(t)$, can be satisfactorily described by a single exponential. In the dissociation of HNO, $P_{cl}(t)$ has a slow and a fast decaying component giving rise to a slow (k_s) and a fast (k_f) rate as seen in the right-hand panel of Fig. 17. This indicates that the phase space is not fully chaotic. However, the difference between k_s and k_f is small, as compared to HCO, for example (left-hand panel of Fig. 17). The slow rate is close to the quantum mechanical average rate, which is not shown in the figure. The average classical rate is almost identical with the classical RRKM rate constant as required for the decay of a micro-canonical ensemble.

The phase space of HCO is much more regular than for the other three systems and that is reflected in a much more pronounced non-exponential classical lifetime distribution function; the difference between the fast rate and the slow rate is considerably larger than for HNO, for example (left-hand panel of Fig. 17). The average classical rate roughly follows the upper bound of the state-specific quantum mechanical rates and is, thus, considerably larger than the quantum mechanical average. Thus, for the four *small* systems described above one can conclude, that the average classical rate is a good approximation of the average quantal rate constant, provided the dynamics is ergodic. For the system, which is not chaotic at all, the agreement becomes less favorable, but at least the order of magnitude is still correct.

However, because the quantum mechanical zero-point energy, E_{zp}, of the products is not conserved in classical mechanics, the situation might be quite different for *large* molecules. Since E_{zp} is available to the dissociation degree of freedom, the classical threshold is smaller than the quantum mechanical one with the effect that the classical rate is approximately shifted by E_{zp} to lower energies compared to the quantum mechanical rate. If the rates rapidly change with energy, as in the threshold region, where the quantum mechanical excess energy is much smaller than E_{zp}, the classical rate might be substantially larger than the quantum mechanical rate. At higher energies, where $E \gg E_{zp}$ and where the increase with energy is more gradual, the non-conservation of E_{zp} becomes less problematic. If the dissociation proceeds over a barrier, quantum mechanical tunneling, not present in classical calculations, might partially compensate — for the wrong reason — this shortcoming in that it lowers the quantum threshold [381] .

For small molecules, E_{zp} is quite small and the non-conservation of zero-point energy does not pose a substantial problem. For example, the quantum mechanical RRKM (i.e., SACM) rates, which include zero-point energy, and the classical RRKM rates, which do not conserve E_{zp}, for HCO and HNO in Fig. 17 are not very different. For large molecules, however, the difference in the two RRKM rates can be significant [382]. Many 'recipes' to circumvent the zero-point energy problem in classical trajectory calculations have been proposed; for a recent discussion see Refs. 383 and 384. Nevertheless, in our opinion a fully satisfactory procedure has not been proposed yet.

At the level of the actual decay curves vs. time the picture is more intricate. As discussed in previous sections, a small molecule decomposes via more or less isolated resonances, each having a particular dissociation rate, and the relationship between the classical and quantum unimolecular dynamics of a micro-canonical ensemble depends on the nature of the classical motion. If the classical dynamics is ergodic and the decomposition intrinsically RRKM, the classical lifetime distribution is exponential and given by Eq. (67). On the other hand, for such a classical system, the quantum dynamics is expected to be statistical state-specific and its survival probability, i.e., the norm of the wave packet in the region of the potential well, as function of time consists of a sum of exponentials for the individual resonance states, each being proportional to $\exp(-k_n t)$. If there are extensive fluctuations in the state-specific rates, the quantum mechanical survival probability will be non-exponential. As the energy increases, the fluctuations become less pronounced and the quantal decay will begin to approach the exponential of the classical decay. The dissociation of NO_2 manifests these rather general trends very clearly (compare Figs. 31 and 43), although the resonances are not completely separated.

The kind of accuracy one can expect from classical calculations for product state distributions has already been addressed in 7.2 and 7.3. If the phase space is irregular and if the fragmentation does not proceed over a barrier, the classical product state distributions satisfactorily approximate the average of the quantum mechanical distributions. Due to interferences — several classical paths may lead to the same final product state — the latter usually show pronounced fluctuations, which purely classical mechanics cannot reproduce. The vibrational (Fig. 37) and rotational (Fig. 42) state distributions of NO following the dissociation of NO_2 illustrate this general behavior. The final state distributions for O_2 calculated for the dissociation of HO_2 are other examples [60]. Because of the zero-point energy problem, at very low excess energies the classical distributions might extend to quantum states, which are not accessible in quantum mechanical calculations, i.e., they may be slightly 'hotter' than the quantum mechanical or experimental distributions. If the dissociation proceeds over a high barrier, as for formaldehyde, the final step, in which the product state distributions are determined, is a very fast and direct process and then classical mechanics is expected to yield very good results [20]. The calculated product state distributions for both, H_2 and CO following the dissociation of H_2CO confirm this view [314, 315, 375].

If a molecule decays in a mode-specific way, the assessment of the accuracy of classical calculations is much more complicated and depends, we believe, sensitively on the initially prepared resonance state. Considering a micro-canonical ensemble certainly will not be appropriate. The initial conditions of the ensemble of trajectories should mimic the quantum mechanical distribution function of coordinates and/or momenta as closely as possible [20, 385]. The gross features of the final state distributions, e.g. the peaking of the CO vibrational distribution in the dissociation of HCO close to the maximum allowed state (Fig. 36), may be qualitatively reproduced. However, more subtle structures are unlikely to be described well, because they often reflect details of the quantum wave function (reflection principle [20]). More work to explore this question is certainly needed.

8.5.2. Non-random excitation — large molecules

The short-time classical unimolecular dynamics for a molecule is in many cases in excellent agreement with quantum dynamics and/or experiment. This holds true both for small molecules and for molecules with many atoms and a large density of states [386, 387]. However, at longer times classical mechanics often fails to retain non-statistical signatures in the intramolecular dynamics, such as vibrational adiabaticity, recurrences, etc., which are present in the quantum dynamics. A nice illustration of this is the comparison of the classical and quantum dynamics of benzene C–H overtone states [388–390]. Classical dynamics describes the initial decay of the overtone state and, thus, gives an accurate linewidth for the overtone absorption transition [388, 389]. However, because classical dynamics does not recover recurrences in the time-evolution of the overtone state, it does not give the detailed structure inside the overtone band. In contrast, the intramolecular dynamics giving rise to the recurrences is accurately described by quantum dynamics and excellent agreement with experiment is obtained [390].

If an intrinsically-RRKM molecule with many atoms is excited non-randomly, its initial classical non-RRKM dynamics may agree with the quantum dynamics for the reasons described above. But at longer times, after a micro-canonical ensemble is created, the classical unimolecular rate constant is much larger than the quantum value, because of the zero-point energy problem. Thus, the short-time unimolecular dynamics of a large molecule will often agree quite well with experiment if the molecule is excited non-randomly. The following is a brief review of two representative studies where the classical dynamics of large molecules agree with experiment and/or quantum dynamics.

The Cl^- ion associates with CH_3Cl to form the $Cl^- \cdots CH_3Cl$ ion-dipole complex. In this association step, only the three intermolecular modes of the complex are excited and energy is transferred to the CH_3Cl intramolecular modes on a much longer timescale. Thus, in the initial decomposition of the complex back to reactants, the complex behaves like a much smaller molecule with only three active degrees of freedom and the initial classical unimolecular rate constant is in excellent agreement with experiment [364]. For longer times, however, there are non-statistical couplings between the intermolecular and intramolecular modes of the complex and the ability of classical mechanics to quantitatively describe these dynamics is unknown.

Trim-ethylene is a moiety with a shallow potential energy well on the reaction path connecting cyclopropane and propylene. Its very short unimolecular lifetime, following different types of initial excitations, has been calculated from classical trajectories [343,344] and compared with both experiment [391] and quantum dynamics [392]. Excellent agreement is found. This is an example of a rather large molecule, for which classical mechanics accurately describes the unimolecular dissociation because of the shallow potential energy minimum and, thus, very short lifetime.

9. OUTLOOK

In this chapter, we discussed the principle quantum mechanical effects inherent to the dynamics of unimolecular dissociation. The starting point of our analysis is the concept of discrete metastable states (resonances) in the dissociation continuum, introduced in Sect. 2 and then amply illustrated in Sects. 5 and 6. Resonances allow one to treat the spectroscopic and kinetic aspects of unimolecular dissociation on equal grounds — they are spectroscopically measurable states and, at the same time, the states in which a molecule can be temporally trapped so that it can be stabilized in collisions with bath particles. The main property of quantum state-resolved unimolecular dissociation is that the lifetimes and hence the dissociation rates strongly fluctuate from state to state — they are intimately related to the shape of the resonance wave functions in the potential well. These fluctuations are universal in that they are observed in mode-specific, statistical state-specific and mixed systems. Thus, the classical notion of an energy dependent reaction rate is not strictly valid in quantum mechanics: Molecules activated with equal amounts of energy but in different resonance states can decay with drastically different rates.

Depending on the nature of the intramolecular dynamics, i.e., regular or irregular, the extent of the fluctuations can be many orders of magnitude.

For NO_2, an irregular system, the rates near the threshold fluctuate over 'only' two orders of magnitude, while for HOCl, a molecule with mainly regular dynamics, the fluctuations of the rates span up to seven orders of magnitude. The (average) dissociation rate predicted by statistical methods (RRKM or SACM, for example) is a smooth function of the excess energy. For irregular molecules, like NO_2, HO_2 or HNO, it agrees — as usually expected — remarkably well with the average quantum mechanical rate. However, even for HCO, which is certainly more regular than irregular, the agreement is acceptable. For HOCl, the most regular system discussed in this chapter, the agreement is less favorable; most of the calculated rates are orders of magnitude smaller than the statistical rate. In the cases of the more irregular systems the distributions of rates are reasonably well described by the prediction of random matrix theory.

Statistical models are widely exploited in kinetic studies and their usefulness certainly does not need any further proof. Nevertheless, the good agreement between average quantum mechanical and statistical rates, calculated for real molecules, as opposed to model systems, and for the same potential energy surfaces, found in the studies over the past decade may convince even the few non-believers.

Of course, the fluctuations are most clearly discernible in species, which are almost completely isolated from the environment. In this concluding section, we will briefly address a more common chemical situation, unimolecular decay in a thermal ensemble, and consider the role of quantum mechanical resonance states in kinetic measurements. From this standpoint, we shall discuss the conditions at which quantum effects are expected to show up, outline new challenges awaiting resonance theory, and point out its further applications.

In a thermal ensemble, the temperature and pressure dependent unimolecular rate constant, $k_{\text{uni}}(\omega, T)$, is directly related to the individual dissociation rates $k_n^{J,K}$ of resonances in a rotating molecule with angular momentum J and its projection K according to

$$k_{\text{uni}}(\omega, T) = \frac{1}{Q_r} \sum_{J=0} (2J+1) \sum_{K=-J}^{J} \sum_{n} e^{-E_n^{J,K}/k_{\text{B}}T} \frac{k_n^{J,K}\,\omega}{k_n^{J,K} + \omega}, \tag{76}$$

where this well known expression [6] assumes the Lindemann mechanism and the limit of strong collisions; here, Q_r is the partition function of the parent molecule, ω is the collision frequency (related to the pressure), k_{B} is the Boltzmann constant and the $E_n^{J,K}$ are the resonance energies. Equation (76) is appropriate for non-overlapping resonances. If the resonances overlap, more general expressions (flux-flux correlation functions) have been derived by Miller [393,394], which in addition to the resonance energies and widths include also the resonance wave functions. Applications of Eq. (76) and its extensions have been done for HCO [395] and HO_2 [396–398], for example.

Even at room temperature, the sum over the angular momentum may include hundreds of rotational states, depending on the rotational constants. Thus, the temperature and pressure dependent unimolecular rate includes averaging over a very large number of resonance states and therefore a reasonable question is to which extent the quantum mechanical fluctuations can survive this averaging. As was shown by Miller [288], the rate $k_{\mathrm{uni}}(\omega, T)$ is actually related to the micro-canonical rate constant averaged over the distribution of resonance widths $Q_\nu(k)$, k_{av}, introduced in Eq. (51),

$$k_{\mathrm{av}}(\omega, E, J, K) = \int_0^\infty dk\, Q_\nu(k; E, J, K)\, \frac{k\,\omega}{k+\omega}\,. \tag{77}$$

Several conclusions can be drawn from Eqs. (76) and (77). First, the influence of fluctuations is the largest when the number of open channels ν is of the order of unity, because then the distribution $Q_\nu(k)$ is the broadest. Second, the effect of a broad distribution of widths is to decrease the observed pressure dependent rate constant as compared to the delta function-like distribution, assumed by statistical theories [288]. The reason is that broad distributions favor small decay rates and the overall dissociation slows down. This trend, pronounced in the fall-of region, was clearly seen in a recent study of thermal rate constants in the unimolecular dissociation of HOCl [399]. The extremely broad distribution of resonances in HOCl caused a decrease by a factor of two in the pressure-dependent rate, as compared to the RRKM predictions. The best chances to see the influence of the quantum mechanical fluctuations on unimolecular rate constants certainly have studies performed close to the dissociation threshold, i.e. at low collision temperatures, because there the distribution of rates is the broadest.

The main obstacle for calculating temperature and pressure dependent rate constants according to Eq. (76) or variants of it is the need for state-resolved dissociation rates for high rotational states. To perform exact quantum mechanical calculations for $J = 40$, for example, is not possible at present time even for a triatomic molecule, especially when it consists of three heavy atoms like O_3. Until now, except for very few studies — HCO and HOCl, for example, discussed in Sect. 5 — most studies of resonance widths have been performed for $J = 0$. However, even at temperatures well below room temperature the molecules with $J = 0$ form only a small fraction of the ensemble. The common way of evaluating the resonance spectra for $J \gg 1$ is the J-shifting approximation [40] discussed in Sect. 4. It has been tested for very low values of J and good agreement with exact calculations has been observed. However, nothing is known about its validity for very large J. One would like to find an independent reliable method of estimating the resonance widths of rotating molecules from calculations at small J's. One way, worth testing in the future, is to combine quantum mechanical and classical calculations, e.g., to treat the vibrational degrees of freedom, which determine the lifetime of the complex, quantum mechanically and to treat the rotational degrees of freedom classically. The problem can be to some extent alleviated by the fact that, as suggested by Eq. (77), for thermal rate constants the distributions of resonance widths are more appropriate quantities to analyze than the individual resonances. Unfortunately, at present we cannot construct these distributions for a real systems without first calculating all resonance states separately.

A candidate for possibly observing quantum mechanical effects in pressure and temperature dependent unimolecular rate constants is ozone. The recombination rates of ozone as measured in the laboratory [400] show pronounced isotope effects, which raise interesting questions concerning the intramolecular dynamics. Actually, ozone is a good aspirant for quantum mechanical effects to become visible in thermal rates, because (a) the density of states is very small at the threshold, (b) the overlap of the near-threshold resonances is not severe, (c) the rates of the individual resonance states greatly fluctuate at low excess energies [42] and (d) many of the states near threshold have wave functions, which are well localized and in many cases even assignable [45]. Future dynamical work, which goes beyond statistical models [401], will show whether or not the isotope effects have anything to do with the fluctuations of the microscopic rates.

A natural question arises as to whether the quantum or at least mode-specific behavior of dissociation can be observed in much larger molecular systems like those discussed in Sect. 8. Deviations of the observed rates from the RRKM model can be used as preliminary indicators of quantum contributions to the reaction. It is necessary, however, that RRKM is not used just as a data fitting device. Recent experimental and theoretical (trajectory) studies of the dissociation dynamics of azo biradicals [402] and multiply charged protein cations [403], for example, suggest that non-ergodic, non-exponential decay can indeed take place in large organic and biomolecules. The non-exponential behavior can be the result of non-ergodic intramolecular dynamics, as discussed in Sect. 8 for the S_N2 reactions, for example, or due to the large fluctuations of the state-dependent rates even for an ergodic molecule, as illustrated for NO_2 in 6.2.

The quantum mechanical description of unimolecular dissociation is a fascinating field. Much has been learned in the past decade, mainly due to the interplay between state-of-the-art experiments and calculations. Nevertheless, many questions are still open and deserve future investigations.

ACKNOWLEDGMENTS

R. Schinke gratefully acknowledges financial support by the Deutsche Forschungsgemeinschaft through the Sonderforschungsbereich 357 "Molekulare Mechanismen unimolekularer Reaktionen". The results of the Göttingen group would have been impossible without the work of many students and postdocs over almost one decade: H.-M. Keller, A. J. Dobbyn, M. Stumpf, C. Beck, J. Hauschildt, J. Weiß, U. Brandt-Pollmann and R. Siebert. S. Yu. Grebenshchikov is grateful to the Alexander von Humboldt-Stiftung for a long-term cooperation fellowship. W. L. Hase gratefully acknowledges financial support from the United States National Science Foundation and thanks the many outstanding students and postdoctoral associates who have worked with him. Finally, the authors thank P. Fleurat-Lessard and W. Keiderling for their invaluable technical help in preparing this chapter.

REFERENCES

[1] P. J. Robinson, K. A. Holbrook, Unimolecular Reactions, Wiley-Interscience, London, 1972.
[2] I. Oref, B. S. Rabinovitch, Acc. Chem. Res. 12 (1979) 166.
[3] K. B. Eisenthal, Acc. Chem. Res. 8 (1975) 118.
[4] G. E. Busch, P. M. Rentzepis, Science 194 (1976) 276.
[5] S. S. Alimpiev, B. G. Sartakov, Laser Chemistry 12 (1992) 147.
[6] T. Baer, W. L. Hase, Unimolecular Reaction Dynamics, Oxford University Press, Oxford, 1996.
[7] W. H. Green, Jr., C. B. Moore, W. F. Polik, Annu. Rev. Phys. Chem. 43 (1992) 591.
[8] R. G. Gilbert, S. C. Smith, Theory of Unimolecular and Recombination Reactions, Blackwell, Oxford, 1990.
[9] D. R. Guyer, W. F. Polik, C. B. Moore, J. Chem. Phys. 84 (1986) 6519.
[10] W. F. Polik, D. R. Guyer, C. B. Moore, J. Chem. Phys. 92 (1990) 3453.
[11] P. M. Felker, A. H. Zewail, J. Chem. Phys. 82 (1985) 2975.
[12] T. Uzer, Phys. Rep. 199 (1991) 73.
[13] M. Gruebele, Adv. Chem. Phys. 114 (2000) 193.
[14] C. Stöck, X. Li, H.-M. Keller, R. Schinke, F. Temps, J. Chem. Phys. 106 (1997) 5333.
[15] H.-M. Keller, T. Schröder, M. Stumpf, C. Stöck, F. Temps, R. Schinke, H.-J. Werner, C. Bauer, P. Rosmus, J. Chem. Phys. 106 (1997) 5359.
[16] J. D. Tobiason, J. R. Dunlop, E. A. Rohlfing, J. Chem. Phys. 103 (1995) 1448.
[17] J. R. Taylor, Scattering Theory: The Quantum Theory on Nonrelativistic Collisions, Wiley, New York, 1972.
[18] L. D. Landau, E. M. Lifschitz, Quantum Mechanics. Non-Relativistic Theory, Pergamon, New York, 1976.

[19] C. Cohen-Tannoudji, B. Diu, F. Laloë, Quantum Mechanics, Vols. I and II, Wiley, New York, 1977.
[20] R. Schinke, Photodissociation Dynamics, Cambridge University Press, Cambridge, 1993.
[21] J. W. Thomas, Jr., J. I. Steinfeld, R. I. McKay, A. E. W. Knight, J. Chem. Phys. 86 (1987) 5909.
[22] Y. S. Choi, C. B. Moore, J. Chem. Phys. 97 (1992) 1010.
[23] A. Geers, J. Kappert, F. Temps, J. Chem. Phys. 101 (1994) 3618.
[24] A. Geers, J. Kappert, F. Temps, J. Chem. Phys. 101 (1994) 3634.
[25] S. Dertinger, A. Geers, J. Kappert, J. Wiebrecht, F. Temps, Faraday Discuss. Chem. Soc. 102 (1995) 31.
[26] G. Dutton, R. J. Barnes, A. Sinha, J. Chem. Phys. 111 (1999) 4976.
[27] A. Callegari, J. Rebstein, R. Jost, T. R. Rizzo, J. Chem. Phys. 111 (1999) 7359.
[28] B. R. Foy, M. P. Casassa, J. C. Stephenson, D. S. King, J. Chem. Phys. 90 (1989) 7037.
[29] F. Reiche, B. Abel, R. D. Beck, T. R. Rizzo, J. Chem. Phys. 112 (2000) 8885.
[30] F. Reiche, B. Abel, R. D. Beck, T. R. Rizzo, J. Chem. Phys. 116 (2002) 10267.
[31] L. R. Khundkar, J. L. Knee, A. H. Zewail, J. Chem. Phys. 87 (1987) 77.
[32] N. Scherer, A. H. Zewail, J. Chem. Phys. 87 (1987) 97.
[33] S. K. Kim, E. R. Lovejoy, C. B. Moore, J. Chem. Phys. 102 (1995) 3202.
[34] S. I. Ionov, G. A. Bruckner, C. Jaques, Y. Chen, C. Wittig, J. Chem. Phys. 99 (1993) 3420.
[35] B. Kirmse, B. Abel, D. Schwarzer, S. Yu. Grebenshchikov, R. Schinke, J. Phys. Chem. A 104 (2000) 10374.
[36] G.-H. Leu, C.-L. Huang, S.-H. Lee, Y.-C. Lee, I.-C. Chen, J. Chem. Phys. 109 (1998) 9340.
[37] A. J. F. Siegert, Phys. Rev. 56 (1939) 750.
[38] V. I. Kukulin, K. M. Krasnopolsky, J. Horáček, Theory of Resonances, Principles and Applications, Kluwer, Dordrecht, 1989.
[39] R. Schinke, C. Beck, S. Yu. Grebenshchikov, H.-M. Keller, Ber. Bunsenges. Phys. Chem. 102 (1998) 593.
[40] J. M. Bowman, J. Phys. Chem. A 102 (1998) 3006.
[41] T. Helgaker, P. Jørgensen, J. Olsen, Molecular Electronic-Structure Theory, Wiley, Chichester, 2000.
[42] R. Siebert, R. Schinke, M. Bittererová, Phys. Chem. Chem. Phys. 3 (2001) 1795.
[43] D. Krankowsky, K. Mauersberger, Science 274 (1996) 1324.
[44] Y. Q. Gao, R. A. Marcus, Science 293 (2001) 259.
[45] R. Siebert, P. Fleurat-Lessard, R. Schinke, M. Bittererová, S. C. Farantos, J. Chem. Phys. 116 (2002) 9749.
[46] J. Z. H. Zhang, Theory and Applications of Quantum Molecular Dynamics, World Scientific, River Edge, NJ, 1996.
[47] R. N. Dixon, J. Chem. Soc. Faraday Trans. 88 (1992) 2575.
[48] D. Wang, J. M. Bowman, J. Chem. Phys. 100 (1994) 1021.
[49] H.-J. Werner, C. Bauer, P. Rosmus, H.-M. Keller, M. Stumpf, R. Schinke, J. Chem. Phys. 102 (1995) 3593.
[50] V. Ryaboy, N. Moiseyev, J. Chem. Phys. 103 (1995) 4061.
[51] H.-M. Keller, H. Flöthmann, A. J. Dobbyn, R. Schinke, H.-J. Werner, C. Bauer, P. Rosmus, J. Chem. Phys. 105 (1996) 4983.
[52] C.-Y. Yang, S. K. Gray, J. Chem. Phys. 107 (1997) 7773.

[53] G. S. Whittier, J. C. Light, J. Chem. Phys. 107 (1997) 1816.
[54] B. Poirier, T. Carrington, Jr., J. Chem. Phys. 116 (2002) 1215.
[55] D. Wang, J. M. Bowman, Chem. Phys. Lett. 235 (1995) 277.
[56] B. K. Kendrick, R. T. Pack, Chem. Phys. Lett. 235 (1995) 291.
[57] B. K. Kendrick, R. T Pack, J. Chem. Phys. 104 (1996) 7502.
[58] J. Dai, J. Z. H. Zhang, J. Chem. Phys. 104 (1996) 3664.
[59] V. A. Mandelshtam, T. P. Grozdanov, H. S. Taylor, J. Chem. Phys. 103 (1995) 10074.
[60] A. J. Dobbyn, M. Stumpf, H.-M. Keller, R. Schinke, J. Chem. Phys. 104 (1996) 8357.
[61] H. Zhang, S. C. Smith, J. Chem. Phys. 115 (2001) 5751.
[62] V. A. Mandelshtam, H. S. Taylor, J. Chem. Soc. Faraday Trans. 93 (1997) 847.
[63] M. Tashiro, S. Kato, Chem. Phys. Lett. 354 (2002) 14.
[64] M. Tashiro, S. Kato, J. Chem. Phys. 117 (2002) 2053.
[65] M. Stumpf, A. J. Dobbyn, D. H. Mordaunt, H.-M. Keller, H. Flöthmann, R. Schinke, H.-J. Werner, K. Yamashita, Faraday Discuss. Chem. Soc. 102 (1995) 193.
[66] S. Skokov, J. M. Bowman, J. Chem. Phys. 110 (1999) 9789.
[67] J. Hauschildt, J. Weiß, C. Beck, S. Yu. Grebenshchikov, R. Düren, R. Schinke, J. Koput, Chem. Phys. Lett. 300 (1999) 569.
[68] J. Weiß, J. Hauschildt, S. Yu. Grebenshchikov, R. Düren, R. Schinke, J. Koput, S. Stamatiadis, S. C. Farantos, J. Chem. Phys. 112 (2000) 77.
[69] J. Weiß, J. Hauschildt, R. Schinke, O. Haan, S. Skokov, J. M. Bowman, V. A. Mandelshtam, K. A. Peterson, J. Chem. Phys. 115 (2001) 8880.
[70] B. Hartke, A. E. Janza, W. Karrlein, J. Manz, V. Mohan, H.-J. Schreier, J. Chem. Phys. 96 (1992) 3569.
[71] S. K. Gray, E. M. Goldfield, J. Phys. Chem. A 105 (2001) 2634.
[72] F. Mrugała, P. Piecuch, V. Širko, O. Bludský, J. Mol. Spec. 555 (2000) 43.
[73] B. Maiti, N. Sathyamurthy, Chem. Phys. Lett. 345 (2001) 461.
[74] M. I. Hernandez, D. C. Clary, J. Chem. Phys. 101 (1994) 2779.
[75] E. M. Goldfield, S. K. Gray, G. C. Schatz, J. Chem. Phys. 102 (1995) 8807.
[76] D. H. Zhang, J. H. Zhang, J. Chem. Phys. 103 (1995) 6512.
[77] T. Yamamoto, S. Kato, J. Chem. Phys. 109 (1998) 9783.
[78] S. Schmatz, P. Botschwina, J. Hauschildt, R. Schinke, J. Chem. Phys. 114 (2001) 5233.
[79] B. Hartke, J. Manz, J. Mathis, Chem. Phys. 139 (1989) 123.
[80] R. H. Bisseling, R. Kosloff, J. Manz, J. Chem. Phys. 83 (1985) 993.
[81] R. H. Bisseling, R. Kosloff, J. Manz, F. Mrugała, J. Römelt, G. Weichselbaumer, J. Chem. Phys. 86 (1987) 2626.
[82] B. Hartke, J. Manz, J. Chem. Phys. 92 (1990) 220.
[83] W. L. Hase, in: W. H. Miller (Ed.), Modern Theoretical Chemistry, Dynamics of Molecular Collisions,Part B, Plenum, New York, 1976.
[84] W. L. Hase, S.-W. Cho, D.-H. Lu, K. N. Swamy, Chem. Phys. 139 (1989) 1.
[85] D. J. Nesbit, Chem. Rev. 88 (1988) 843.
[86] J. M. Hutson, Annu. Rev. Phys. Chem. 41 (1990) 123.
[87] Z. Bačić, R. E. Miller, J. Phys. Chem. 100 (1996) 12945.
[88] A. Rohrbacher, N. Halberstadt, K. C. Janda, Annu. Rev. Phys. Chem. 51 (2000) 405.
[89] L. Oudejans, R. E. Miller, Annu. Rev. Phys. Chem. 52 (2001) 607.
[90] N. Halberstadt, S. Serna, O. Roncero, K. C. Janda, J. Chem. Phys. 97 (1992) 341.

[91] O. Roncero, S. K. Gray, J. Chem. Phys. 104 (1996) 4999.
[92] K. Liu, R. T. Skodje, D. E. Manolopoulos, PhysChemCom 5 (2002) 27.
[93] C. Mahaux, H. A. Weidenmüller, Shell-Model Approach to Nuclear Reactions, North-Holland, Amsterdam, 1969.
[94] T. F. Gallagher, Rydberg Atoms, Cambridge University Press, Cambridge, 1994.
[95] J.-M. Rost, Phys. Rep. 297 (1998) 271.
[96] W. Domcke, Phys. Rep. 208 (1991) 97.
[97] D. M. Neumark, Ann. Rev. Phys. Chem. 43 (1992) 153.
[98] H. Hoinkes, Rev. Mod. Phys. 52 (1980) 933.
[99] D. G. Truhlar (Ed.), Resonances in Electron-Molecule Scattering, van der Waals Complexes, and Reactive Chemical Dynamics, American Chemical Society, Washington, D.C., 1984.
[100] J. A. Beswick, J. Jortner, Adv. Chem. Phys. 47 (1981) 363.
[101] S. Alavi, T. Seideman, J. Chem. Phys. 115 (2001) 1882.
[102] R. Schinke, H.-M. Keller, M. Stumpf, A. J. Dobbyn, J. Phys. B 28 (1995) 3081.
[103] M. S. Child, Molecular Collision Theory, Academic Press, London, 1974.
[104] G. Gamow, Constitution of Atomic Nuclei and Radioactivity, Oxford University Press, Oxford, 1931.
[105] P. Kapur, R Peierls, Proc. Roy. Soc. Lond. A 166 (1938) 277.
[106] J. Schwinger, Ann. Phys. N.Y. 9 (1960) 169.
[107] M. Reed, B. Simon, Methods of Modern Mathematical Physics, Vol. 4, Academic Press, New York, 1978.
[108] A. M. Perelomov, Ya. B. Zel'dovich, Quantum Mechanics, World Sientific, Singapore, 1998.
[109] R. M. More, E. Gerjuoy, Phys. Rev. A 7 (1973) 1289.
[110] S. Yu. Grebenshchikov, J. Phys. Chem. A 104 (2000) 10409.
[111] G.Breit, E. P. Wigner, Phys. Rev. 49 (1936) 519,642.
[112] U. Fano, Phys. Rev. 124 (1961) 1866.
[113] O. I. Tolstikhin, V. N. Ostrovsky, H. Nakamura, Phys. Rev. A 58 (1998) 2077.
[114] M. C. Gutzwiller, Chaos in Classical and Quantum Mechanics, Springer, New York, 1990.
[115] D.-h. Lu, W. L. Hase, J. Chem. Phys. 90 (1989) 1557.
[116] R. Hernandez, W. H. Miller, C. B. Moore, W. F. Polik, J. Chem. Phys. 99 (1993) 950.
[117] K. Someda, H. Nakamura, F. H. Mies, Chem. Phys. 187 (1994) 195.
[118] T. Ericson, T. Mayer-Kuckuk, Ann. Rev. Nuc. Sci. 16 (1966) 183.
[119] F. H. Mies, M. Krauss, J. Chem. Phys. 45 (1966) 4455.
[120] Y. V. Fyodorov, H. J. Sommers, J. Math. Phys. 38 (1997) 1918.
[121] Y. Alhassid, Y. V. Fyodorov, J. Phys. Chem. A 102 (1998) 9577.
[122] H. Lefebvre-Brion, R. W. Field, Perturbations in the Spectra of Diatomic Molecules, Academic Press, Orlando, 1986.
[123] M. Philipp, P. von Brentano, G. Pascovici, A. Richter, Phys. Rev. E 62 (2000) 1922.
[124] C. Dembowski, B. Dietz, H. D. Gräf, H. L. Harney, W. D. Heiss, H. Rehfeld, A. Richter, Phys. Rev. Lett. 86 (2001) 787.
[125] E. Persson, K. Pichugin, I. Rotter, P. Šeba, Phys. Rev. E 58 (1998) 8001.
[126] A. Callegari, T. R. Rizzo, Chem. Soc. Rev. 30 (2001) 214.
[127] U. Brandt-Pollmann, J. Weiß, R. Schinke, J. Chem. Phys. 115 (2001) 8876.
[128] J. Weiß, R. Schinke, J. Chem. Phys. 115 (2001) 3173.
[129] G. Hirsch, R. J. Buenker, C. Petrongolo, Mol. Phys. 73 (1991) 1085.

[130] S. Mahapatra, H. Köppel, L. S. Cederbaum, P. Stampfuß, W. Wenzel, Chem. Phys. 259 (2000) 211.
[131] J. Miyawaki, K. Yamanouchi, S. Tsuchiya, Chem. Phys. Lett. 180 (1991) 287.
[132] U. Robra, H. Zacharias, K. Welge, Z. Phys. D 16 (1990) 175.
[133] S. A. Reid, H. Reisler, J. Phys. Chem. 100 (1996) 474.
[134] J. Miyawaki, K. Yamanouchi, S. Tsuchiya, J. Chem. Phys. 99 (1993) 254.
[135] S. A. Reid, J. T. Brandon, M. Hunter, H. Reisler, J. Chem. Phys. 99 (1993) 4860.
[136] S. A. Reid, D. C. Robie, H. Reisler, J. Chem. Phys. 100 (1994) 4256.
[137] B. Abel, H. H. Hamann, N. Lange, Faraday Discuss. Chem. Soc. 102 (1995) 147.
[138] P. M. Felker, A. H. Zewail, Adv. Chem. Phys. 70 (1988) 265.
[139] A. H. Zewail, Faraday Discuss. Chem. Soc. 91 (1991) 207.
[140] A. H. Zewail, in: J. Manz, L. Wöste (Eds.), Femtosecond Chemistry, Verlag Chemie, Weinheim, 1995.
[141] F. F. Crim, in: M. N. R. Ashfold, J. E. Baggott (Eds.), Molecular Photodissociation Dynamics, The Royal Society of Chemistry, London, 1987.
[142] F. F. Crim, Ann. Rev. Phys. Chem. 44 (1993) 397.
[143] T. R. Rizzo, C. C. Hayden, F. F. Crim, J. Chem. Phys. 81 (1984) 4501.
[144] T. M. Ticich, T. R. Rizzo, H. R. Dubal, F. F. Crim, J. Chem. Phys. 84 (1986) 1508.
[145] X. Luo, T. R. Rizzo, J. Chem. Phys. 94 (1991) 889.
[146] X. Luo, T. R. Rizzo, J. Chem. Phys. 96 (1992) 5129.
[147] H.-L. Dai, R. W. Field (Eds.), Molecular Dynamics and Spectroscopy by Stimulated Emission Pumping, World Scientific, Singapore, 1995.
[148] W. A. Lester, Jr., in: W. H. Miller (Ed.), Dynamics of Molecular Collisions, Plenum Press, New York, 1976.
[149] S.-W. Cho, A. F. Wagner, B. Gazdy, J. M. Bowman, J. Chem. Phys. 96 (1992) 2799.
[150] S.-W. Cho, A. F. Wagner, B. Gazdy, J. M. Bowman, J. Chem. Phys. 96 (1992) 2812.
[151] E. P. Wigner, Phys. Rev. 98 (1955) 145.
[152] F. T. Smith, Phys. Rev. 118 (1960) 349.
[153] D. J. Kouri, M. Arnold, D. K. Hoffman, Chem. Phys. Lett. 203 (1993) 166.
[154] D. J. Kouri, Y. Huang, W. Zhu, D. K. Hoffman, J. Chem. Phys. 100 (1994) 3662.
[155] D. K. Hoffman, Y. Huang, W. Zhu, D. J. Kouri, J. Chem. Phys. 101 (1994) 1242.
[156] H. W. Jang, J. Light, J. Chem. Phys. 102 (1995) 3262.
[157] D. E. Manolopuolos, R. E. Wyatt, Chem. Phys. Lett. 152 (1988) 23.
[158] D. E. Manolopuolos, M. D'Mello, R. E. Wyatt, J. Chem. Phys. 91 (1989) 6096.
[159] D. E. Manolopuolos, M. D'Mello, R. E. Wyatt, J. Chem. Phys. 93 (1990) 403.
[160] H.-M. Keller, R. Schinke, J. Chem. Phys. 110 (1999) 9887.
[161] W. Kohn, Phys. Rev. 74 (1948) 1763.
[162] E. P. Wigner, L. Eisenbud, Phys. Rev. 72 (1947) 29.
[163] E. J. Heller, Acc. Chem. Res. 14 (1981) 368.
[164] R. Kosloff, D. Kosloff, J. Comput. Phys. 63 (1986) 363.
[165] V. Mohan, N. Sathyamurthy, Comp. Phys. Rep. 7 (1988) 213.
[166] R. Kosloff, Annu. Rev. Phys. Chem. 45 (1994) 145.
[167] J. Manz, in: V. Sundström (Ed.), Femtochemistry and Femtobiology: Ultrafast Dynamics at Atomic Scale Resolution, Imperial College, London, 1997.
[168] W. Domcke, G. Stock, Adv. Chem. Phys. 100 (1997) 1.
[169] S. K. Gray, J. Chem. Phys. 96 (1992) 6543.
[170] A. J. H. M. Meijer, E. M. Goldfield, J. Chem. Phys. 110 (1999) 870.

[171] H. Tal-Ezer, R. Kosloff, J. Chem. Phys. 81 (1984) 3967.
[172] Y. Huang, D. J. Kouri, D. K. Hoffman, Chem. Phys. Lett. 225 (1994) 37.
[173] T. J. Park, J. C. Light, J. Chem. Phys. 85 (1986) 5870.
[174] N. Balakrishnan, C. Kalyanaraman, N. Sathyamurthy, Phys. Rep. 280 (1997) 79.
[175] R. B. Gerber, R. Kosloff, M. Berman, Computer Physics Reports 5 (1986) 59.
[176] G. C. Groenenboom, D. T. Colbert, J. Chem. Phys. 99 (1993) 9681.
[177] D. Neuhauser, M. Baer, J. Chem. Phys. 90 (1989) 4351.
[178] G. R. B. Prony, J. Ec. Polytech. (Paris) 1 (1795) 22.
[179] S. L. Marple, Jr., Digital Spectral Analysis with Applications, Prentice Hall, Englewood Cliffs, NJ, 1987.
[180] R. Roy, B. G. Sumpter, G. A. Pfeffer, S. K. Gray, D. W. Noid, Phys. Rep. 205 (1991) 109.
[181] M. R. Wall, D. Neuhauser, J. Chem. Phys. 102 (1995) 8011.
[182] V. A. Mandelshtam, H. S. Taylor, Phys. Rev. Lett. 78 (1997) 3274.
[183] V. A. Mandelshtam, H. S. Taylor, J. Chem. Phys. 107 (1997) 6756.
[184] R. Chen, H. Guo, Chem. Phys. Lett. 279 (1997) 252.
[185] H.-G. Yu, S. C. Smith, Chem. Phys. Lett. 283 (1998) 69.
[186] V. A. Mandelshtam, J. Phys. Chem. A 105 (2001) 2764.
[187] O. I. Tolstikhin, V. N. Ostrovsky, H. Nakamura, Phys. Rev. Lett. 79 (1997) 2026.
[188] A. U. Hazi, H. S. Taylor, Phys. Rev. A 1 (1970) 1109.
[189] H. S. Taylor, Adv. Chem. Phys. 18 (1970) 91.
[190] V. A. Mandelshtam, T. R. Ravuri, H. S. Taylor, Phys. Rev. Lett. 70 (1993) 1932.
[191] J. M. Bowman, B. Gazdy, S.-W. Cho, A. F. Wagner, J. Chem. Phys. 94 (1991) 4192.
[192] W. P. Reinhardt, Ann. Rev. Phys. Chem. 33 (1982) 223.
[193] N. Moiseyev, Phys. Rep. 302 (1998) 211.
[194] J. Aguilar, J. M. Combes, Commun. Math. Phys. 22 (1971) 265.
[195] E. Baslev, J. M. Combes, Commun. Math. Phys. 22 (1971) 280.
[196] B. Simon, Commun. Math. Phys. 27 (1972) 1.
[197] J. N. L. Connor, A. D. Smith, J. Chem. Phys. 78 (1983) 6161.
[198] G. Jolicard, E. J. Austin, Chem. Phys. 103 (1986) 295.
[199] G. Jolicard, C. Leforestier, E. J. Austin, J. Chem. Phys. 88 (1988) 1026.
[200] I. Last, D. Neuhauser, M. Baer, J. Chem. Phys. 96 (1992) 2017.
[201] I. Last, A. Baram, M. Baer, Chem. Phys. Lett. 195 (1992) 435.
[202] T. Seideman, W. H. Miller, J. Chem. Phys. 96 (1992) 4412.
[203] U. V. Riss, H.-D. Meyer, J. Phys. B 24 (1993) 4503.
[204] H. Feshbach, Ann. Phys. 19 (1962) 287.
[205] V. V. Sokolov, V. G. Zelevinsky, Nucl. Phys. A 504 (1988) 562.
[206] V. V. Sokolov, V. G. Zelevinsky, Ann. Phys. (N.Y.) 216 (1992) 323.
[207] U. V. Riss, H.-D. Meyer, J. Phys. B 28 (1995) 1475.
[208] T. P. Grozdanov, V. A. Mandelshtam, H. S. Taylor, J. Chem. Phys. 103 (1995) 7990.
[209] J. C. Light, T. Carrington, Adv. Chem. Phys. 114 (2000) 263.
[210] Z. Bačić, J. C. Light, Ann. Rev. Phys. Chem. 40 (1989) 469.
[211] C. Cerjan (Ed.), Numerical Grid Methods and Their Applications to Schrödinger's Equation, Kluwer Academic Publishers, Dordrecht, 1993.
[212] C. Leforestier, K. Yamashita, N. Moiseyev, J. Chem. Phys. 103 (1995) 8468.
[213] G.-J. Kroes, D. Neuhauser, J. Chem. Phys. 105 (1996) 9104.
[214] H.-G. Yu, G. Nyman, Chem. Phys. Lett. 298 (1998) 27.

[215] Y. Huang, W. Zhu, D. J. Kouri, D. K. Hoffman, Chem. Phys. Lett. 206 (1993) 96.
[216] D. Neuhauser, J. Chem. Phys. 93 (1990) 2611.
[217] D. Neuhauser, J. Chem. Phys. 95 (1991) 4927.
[218] D. Neuhauser, J. Chem. Phys. 100 (1994) 5076.
[219] D. Xie, R. Chen, H. Guo, J. Chem. Phys. 112 (2000) 5263.
[220] J. K. Cullum, R. A. Willoughby, Lanczos Algorithms for Large Symmetric Eigenvalue Computations, Birkhauser, Boston, 1985.
[221] V. A. Mandelshtam, H. S. Taylor, J. Chem. Phys. 102 (1995) 7390.
[222] J. C. Light, J. P. Hamilton, J. V. Lill, J. Chem. Phys. 82 (1985) 1900.
[223] D. T. Colbert, W. H. Miller, J. Chem. Phys. 96 (1992) 1982.
[224] C. Leforestier, J. Chem. Phys. 94 (1991) 6388.
[225] F. A. Gianturco, The Transfer of Molecular Energy by Collisions, Springer, Heidelberg, 1979.
[226] J. M. Bowman, Chem. Phys. Lett. 217 (1994) 36.
[227] S. Skokov, J. M. Bowman, J. Chem. Phys. 111 (1999) 4933.
[228] J. Zhang, S. I. Chu, W. H. Miller, J. Chem. Phys. 88 (1988) 6233.
[229] J. Zhang, W. H. Miller, J. Chem. Phys. 92 (1990) 1811.
[230] B. R. Johnson, J. Comput. Phys. 13 (1973) 445.
[231] A. Messiah, Quantum Mechanics, Vol. I and II, North-Holland Publishing Company, Amsterdam, 1972.
[232] Yu. N. Demkov, Variational Principles in the Theory of Collisions, Pergamon Press, Oxford, 1963.
[233] R. K. Nesbet, Variational Methods in Electron-Atom Scattering Theory, Plenum, New York, 1980.
[234] L. D. Landau, E. M. Lifshitz, Mechanics, Pergamon, New York, 1976.
[235] B. Hartke, R. Kosloff, S. Ruhman, Chem. Phys. Lett. 158 (1989) 238.
[236] M. Abramowitz, I. A. Stegun (Ed.), Handbook of Mathematical Functions, Dover, New York, 1970.
[237] U. V. Riss, H.-D. Meyer, J. Chem. Phys. 105 (1996) 1409.
[238] D. W. Neyer, P. L. Houston, in: K. Liu, A. Wagner (Eds.), The Chemical Dynamics and Kinetics of Small Radicals, World Scientific, Singapore, 1994.
[239] D. J. Hucknall, Chemistry of Hydrocarbon Combustion, Chapman and Hall, London, 1985.
[240] B. J. Finlayson-Pitts, J. N. Pitts, Jr.., Atmospheric Chemistry, Wiley, New York, 1986.
[241] K. Tanaka, E. R. Davidson, J. Chem. Phys. 70 (1979) 2904.
[242] G. W. Adams, X. Zhao, R. W. Field, J. Mol. Spectrosc. 160 (1993) 11.
[243] D. W. Neyer, X. Luo, P. L. Houston, I. Burak, J. Chem. Phys. 98 (1993) 5095.
[244] D. W. Neyer, X. Luo, I. Burak, P. L. Houston, J. Chem. Phys. 102 (1995) 1645.
[245] J. M. Bowman, J. S. Bittman, L. B. Harding, J. Chem. Phys. 85 (1986) 911.
[246] J. M. Bowman, B. Gazdy, J. Chem. Phys. 94 (1991) 816.
[247] J. M. Bowman, B. Gazdy, J. Chem. Phys. 95 (1991) 6309.
[248] J. M. Bowman, B. Gazdy, Chem. Phys. Lett. 200 (1992) 311.
[249] W. L. Hase, Acc. Chem. Res. 16 (1983) 258.
[250] M. Quack, J. Troe, Ber. Bunsenges. Phys. Chem. 78 (1974) 240.
[251] M. Quack, J. Troe, Ber. Bunsenges. Phys. Chem. 79 (1975) 170.
[252] J. D. Tobiason, J. R. Dunlop, E. A. Rohlfing, Chem. Phys. Lett. 235 (1995) 268.
[253] A. Troellsch, F. Temps, Z. Phys. Chem. 215 (2001) 207.
[254] S. Skokov, K. A. Peterson, J. M. Bowman, J. Chem. Phys. 109 (1998) 2662.

[255] K. A. Peterson, S. Skokov, J. M. Bowman, J. Chem. Phys. 111 (1999) 7446.
[256] S. Skokov, J. M. Bowman, V. A. Mandelshtam, Phys. Chem. Chem. Phys. 1 (1999) 1279.
[257] R. Jost, M. Joyeux, S. Skokov, J. M. Bowman, J. Chem. Phys. 111 (1999) 6807.
[258] M. Joyeux, S. C. Farantos, R. Schinke, J. Phys. Chem. A 106 (2002) 5407.
[259] B. Abel, A. Charvát, S. F. Deppe, H. H. Hamann, Ber. Bunsenges. Phys. Chem. 101 (1997) 329.
[260] A. Charvát, S. F. Deppe, H. H. Hamann, B. Abel, J. Mol. Spec. 185 (1997) 336.
[261] H. H. Hamann, A. Charvát, B. Abel, S. A. Kovalenko, A. A. Kachanov, J. Chem. Phys. 106 (1997) 3103.
[262] S. Michielsen, A. J. Merer, S. A. Rice, F. A. Novak, K. F. Freed, Y. Hamada, J. Chem. Phys. 74 (1981) 3089.
[263] A. Callegari, J. Rebstein, J. S. Muenter, R. Jost, T. R. Rizzo, J. Chem. Phys. 111 (1999) 123.
[264] S. Skokov, K. A. Peterson, J. M. Bowman, Chem. Phys. Lett. 312 (1999) 494.
[265] A. Callegari, R. Schmied, P. Theulé, J. Rebstein, T. R. Rizzo, Phys. Chem. Chem. Phys. 3 (2001) 2245.
[266] J. Pearson, A. J. Orr-Ewing, M. N. R. Ashfold, R. N. Dixon, J. Chem. Phys. 106 (1997) 5850.
[267] W. F. Polik, C. B. Moore, W. H. Miller, J. Chem. Phys. 89 (1988) 3584.
[268] H. Hippler, M. Siefke, H. Stark, J. Troe, Phys. Chem. Chem. Phys. 1 (1999) 57.
[269] L. B. Harding, H. Stark, J. Troe, V. G. Ushakov, Phys. Chem. Chem. Phys. 1 (1999) 63.
[270] D. Reignier, T. Stoecklin, P. Halvick, A. Voronin, J. C. Rayez, Phys. Chem. Chem. Phys. 3 (2001) 2726.
[271] S. Yu. Grebenshchikov, C. Beck, H. Flöthmann, R. Schinke, S. Kato, J. Chem. Phys. 111 (1999) 619.
[272] E. Haller, H. Köppel, L. S. Cederbaum, J. Mol. Spec. 111 (1985) 377.
[273] O. L. A. Monti, H. Dickinson, S. R. Mackenzie, T. P. Softley, J. Chem. Phys. 112 (2000) 3699.
[274] D. V. Savin, V. V. Sokolov, Phys. Rev. E 56 (1997) R4911.
[275] I. Bezel, D. Stolyarov, C. Wittig, J. Phys. Chem. A 103 (1999) 10268.
[276] I. Bezel, P. Ionov, C. Wittig, J. Chem. Phys. 111 (1999) 9267.
[277] S.Yu. Grebenshchikov, H. Flöthmann, R. Schinke, I. Bezel, C. Wittig, S. Kato, Chem. Phys. Lett. 285 (1998) 410.
[278] B. Abel, B. Kirmse, J. Troe, D. Schwarzer, J. Chem. Phys. 115 (2001) 6522.
[279] B. Abel, N. Lange, J. Troe, J. Chem. Phys. 115 (2001) 6531.
[280] C. E. Porter, R. G. Thomas, Phys. Rev. 104 (1956) 483.
[281] C. E. Porter, Statistical Theory of Spectra: Fluctuations, Academic Press, New York, 1965.
[282] T. A. Brody, J. Flores, J. B. French, P. A. Mello, A. Pandey, S. S. M. Wong, Rev. Mod. Phys. 53 (1981) 385.
[283] M. Carmeli, Statistical Theory and Random Matrices, Dekker, New York, 1983.
[284] E. B. Stechel, E. J. Heller, Annu. Rev. Phys. Chem. 35 (1984) 563.
[285] R. D. Levine, Adv. Chem. Phys. 70, Part 1 (1988) 53.
[286] T. Zimmermann, L. S. Cederbaum, H.-D. Meyer, H. Köppel, J. Phys. Chem. 91 (1987) 4446.
[287] W. F. Polik, D. R. Guyer, W. H. Miller, C. B. Moore, J. Chem. Phys. 92 (1990) 3471.

[288] W. H. Miller, J. Phys. Chem. 92 (1988) 4261.
[289] E. R. Lovejoy, S. K. Kim, C. B. Moore, Science 256 (1992) 1541.
[290] I.-C. Chen, C. B. Moore, J. Phys. Chem. 94 (1990) 263.
[291] W. D. Allen, H. F. Schaefer III, J. Chem. Phys. 89 (1988) 329.
[292] I. W. M. Smith, Nature 358 (1992) 279.
[293] R. A. Marcus, Science 256 (1992) 1523.
[294] S. I. Ionov, H. F. Davis, K. Mikhaylichenko, L. Valachovic, R. A. Beaudet, C. Wittig, J. Chem. Phys. 101 (1994) 4809.
[295] J. D. Gezelter, W. H. Miller, J. Chem. Phys. 104 (1996) 3546.
[296] R. A. King, W. D. Allen, B. Ma, H. F. Schaefer III, Faraday Discuss. 110 (1998) 23.
[297] R. A. King, W. D. Allen, H. F. Schaefer III, J. Chem. Phys. 112 (2000) 5585.
[298] P. Andresen, G. S. Ondrey, B. Titze, E. W. Rothe, J. Chem. Phys. 80 (1984) 2548.
[299] A. Sanov, S. A. Reid, H. Reisler, J. Phys. Chem. 99 (1995) 13637.
[300] J. C. Stephenson, M. P. Casassa, D. S. King, J. Chem. Phys. 89 (1988) 1378.
[301] T. J. Butenhoff, K. L. Carleton, C. B. Moore, J. Chem. Phys. 92 (1990) 377.
[302] M. L. Costen, H. Katayanagi, G. E. Hall, J. Phys. Chem. A 104 (2000) 10247.
[303] R. N. Zare, Angular Momentum, Wiley, New York, 1988.
[304] G. E. Hall, P. L. Houston, Annu. Rev. Phys. Chem. 40 (1989) 375.
[305] R. Bersohn, J. Phys. Chem. 88 (1984) 5145.
[306] H. Sato, Chem. Rev. 101 (2001) 2686.
[307] G. G. Balint-Kurti, M. Shapiro, Chem. Phys. 61 (1981) 137.
[308] G. G. Balint-Kurti, R. N. Dixon, C. C. Marston, J. Chem. Soc. Faraday Trans. 86 (1990) 1741.
[309] P. Pechukas, J. C. Light, C. Rankin, J. Chem. Phys. 44 (1966) 794.
[310] J. A. Harrison, X. Yang, M. Rösslein, P. Felder, J. R. Huber, J. Phys. Chem. 98 (1994) 12260.
[311] H. Reisler, H.-M. Keller, R. Schinke, Comm. At. Mol. Phys. 30 (1994) 191.
[312] D. J. Bamford, S. V. Filseth, M. F. Foltz, J. W. Hepburn, C. B. Moore, J. Chem. Phys. 82 (1985) 3032.
[313] J. D. Goddard, H. F. Schaefer III, J. Chem. Phys. 70 (1979) 5117.
[314] R. Schinke, Chem. Phys. Lett. 120 (1985) 129.
[315] Y.-T. Chang, C. Minichino, W. H. Miller, J. Chem. Phys. 96 (1992) 4341.
[316] R. Schinke, H. Meyer, U. Buck, G. H. F. Diercksen, J. Chem. Phys. 80 (1984) 5518.
[317] H.-M. Keller, R. Schinke, J. Chem. Soc., Faraday Trans. 93 (1997) 879.
[318] R. Schinke, V. Engel, P. Andresen, D. Häusler, G. G. Balint-Kurti, Phys. Rev. Lett. 55 (1985) 1180.
[319] P. Pechukas, J. C. Light, J. Chem. Phys. 42 (1965) 3285.
[320] L. Bonnet, J. C. Rayez, J. Chem. Phys. 110 (1999) 4772.
[321] L. Bonnet, J. C. Rayez, Phys. Chem. Chem. Phys. 1 (1999) 2383.
[322] P. Larregaray, L. Bonnet, J. C. Rayez, J. Chem. Phys. 114 (2001) 3349.
[323] D. L. Bunker, J. Chem. Phys. 37 (1962) 393.
[324] D. L. Bunker, J. Chem. Phys. 40 (1964) 1946.
[325] K. Bolton, W. L. Hase, G. H. Peslherbe, in: D. L. Thompson (Ed.), Multidimensional Molecular Dynamics Methods, World Scientific, London, 1998.
[326] J. M. Haile, Molecular Dynamics Simulations, Wiley, New York, 1992.
[327] A. J. Lichtenberg, M. A. Lieberman, Regular and Stochastic Motion, Springer-Verlag, New-York, 1983.
[328] R. J. Wolf, W. L. Hase, J. Chem. Phys. 72 (1980) 316.

[329] G. H. Peslherbe, H. Wang, W. L. Hase, Adv. Chem. Phys. 105 (1999) 171.
[330] C. S. Sloan, W. L. Hase, J. Chem. Phys. 66 (1977) 1523.
[331] D. A. McQuarrie, Statistical Thermodynamics, Harper, New York, 1973.
[332] D. L. Bunker, W. L. Hase, J. Chem. Phys. 59 (1973) 4673.
[333] N. B. Slater, Theory of Unimolecular Reaction Rates, Cornell University, Ithaca, 1959.
[334] W. H. Miller, J. Chem. Phys. 65 (1976) 2216.
[335] P. Pechukas, in: W. H. Miller (Ed.), Dynamics of Molecular Collisions, Part B, Plenum Press, New York, 1976.
[336] J. D. Doll, J. Chem. Phys. 73 (1980) 2760.
[337] D. Chandler, Introduction to Modern Statistical Mechanics, Oxford University Press, New York, 1987.
[338] W. L. Hase, D. G. Buckowski, K. N. Swamy, J. Phys. Chem. 87 (1983) 2754.
[339] G. H. Peslherbe, W. L. Hase, J. Chem. Phys. 105 (1996) 7432.
[340] X. Hu, W. L. Hase, J. Chem. Phys. 95 (1991) 8073.
[341] S. R. Vande Linde, W. L. Hase, J. Chem. Phys. 93 (1990) 7962.
[342] D. V. Shalashilin, D. L. Thompson, J. Chem. Phys. 105 (1996) 1833.
[343] C. Doubleday, Jr., K. Bolton, G. H. Peslherbe, W. L. Hase, J. Am. Chem. Soc. 118 (1996) 9922.
[344] C. Doubleday, Jr., K. Bolton, W. L. Hase, J. Phys. Chem. A 102 (1998) 3648.
[345] D. V. Shalashilin, D. L. Thompson, J. Chem. Phys. 107 (1997) 6204.
[346] D. V. Shalashilin, D. L. Thompson, J. Phys. Chem. 101 (1997) 961.
[347] D. J. Wales, J. P. K. Doye, M. A. Miller, P. N. Mortensen, , T. R. Walsh, Adv. Chem. Phys. 115 (2000) 1.
[348] D. J. Wales, private communication .
[349] M. Tabor, Chaos and Integrability in Nonlinear Dynamics, Wiley, New York, 1989.
[350] R. Blümel, W. P. Reinhardt, Chaos in Atomic Physics, Cambridge University Press, Cambridge, 1997.
[351] M. Joyeux, D. Sugny, M. Lombardi, R. Jost, R. Schinke, S. Skokov, J. M. Bowman, J. Chem. Phys. 113 (2000) 9610.
[352] D. W. Noid, M. L. Koszykowski, R. A. Marcus, J. Chem. Phys. 67 (1977) 404.
[353] G. Benettin, L. Galgani, J. M. Strelcyn, Phys. Rev. A 14 (1976) 2338.
[354] H. Wang, W. L. Hase, Int. J. Mass Spectrom. Ion Process. 167/168 (1997) 573.
[355] R. B. Shirts, W. P. Reinhardt, J. Chem. Phys. 77 (1982) 5204.
[356] M. J. Davis, Int. Rev. Phys. Chem. 14 (1995) 15.
[357] V. I. Arnold, Dokl. Akad. Nauk. SSSR 156 (1964) 9.
[358] G. S. Ezra, in: W. L. Hase (Ed.), Advances in Classical Trajectory Methods, Vol. 1, JAI, London, 1992.
[359] S. Keshavamurthy, G. S. Ezra, J. Chem. Phys. 107 (1997) 156.
[360] M. J. Davis, J. Chem. Phys. 83 (1985) 1016.
[361] M. J. Davis, S. K. Gray, J. Chem. Phys. 84 (1986) 5389.
[362] W. L. Hase, Science 266 (1994) 998.
[363] W. L. Hase, Acc. Chem. Res. 31 (1998) 659.
[364] G. H. Peslherbe, H. Wang, W. L. Hase, J. Chem. Phys. 102 (1995) 5626.
[365] A. A. Viggiano, R. A. Morris, J. S. Paschewitz, J. F. Paulson, J. Am. Chem. Soc. 114 (1992) 10477.
[366] K. Song, W. L. Hase, L. Sun, J. Am. Chem. Soc. 123 (2001) 5753.
[367] Y. J. Cho, S. R. Vande Linde, L. Zhu, W. L. Hase, J. Chem. Phys. 96 (1992) 8275.
[368] R. Car, R. Parrinello, Phys. Rev. Lett. 55 (1985) 2471.

[369] H. Tachikawa, M. Igarashi, Chem. Phys. Lett. 303 (1999) 81.
[370] H. Tachikawa, J. Phys. Chem. A 105 (2001) 1260.
[371] S. Raugei, G. Cardini, , V. Schettino, J. Chem. Phys. 111 (1999) 10887.
[372] S. Raugei, G. Cardini, V. Schettino, J. Chem. Phys. 114 (2001) 4089.
[373] M. Pagliai, S. Raugei, G. Cardini, , V. Schettino, Phys. Chem. Chem. Phys. 3 (2001) 2559.
[374] K. Bolton, H. B. Schlegel, W. L. Hase, K. Song, Phys. Chem. Chem. Phys. 1 (1999) 999.
[375] W. Chen, W. L. Hase, , H. B. Schlegel, Chem. Phys. Lett. 228 (1994) 436.
[376] K. M. Forsythe, S. K. Gray, S. J. Klippenstein, G. E. Hall, J. Chem. Phys. 115 (2001) 2134.
[377] A. Warshel, M. Levitt, J. Mol. Biol. 103 (1976) 227.
[378] S. O. Meroueh, Y. F. Wang, W. L. Hase, J. Phys. Chem. A 106 (2002) 9983.
[379] G. C. Schatz, in: W. L. Hase (Ed.), Advances in Classical Trajectory Methods, Vol. 3, JAI Press, London, 1998.
[380] W. H. Miller, J. Phys. Chem. A 105 (2001) 2942.
[381] K. N. Swamy, W. L. Hase, J. Phys. Chem. 87 (1983) 2754.
[382] W. L. Hase, D. G. Buckowski, J. Comput. Chem. 3 (1982) 335.
[383] U. Müller, G. Stock, J. Chem. Phys. 111 (1999) 77.
[384] G. Stock, U. Müller, J. Chem. Phys. 111 (1999) 65.
[385] W. Schleich, Quantum Optics in Phase Space, Wiley-VCH, Berlin, 2001.
[386] E. J. Heller, J. Chem. Phys. 62 (1975) 1544.
[387] J. Brickmann, R. Pfeiffer, P. C. Schmidt, Ber. Bunsenges. Phys. Chem. 88 (1984) 382.
[388] E. L. Sibert, III, J. T. Hynes, W. P. Reinhardt, J. Chem. Phys. 81 (1984) 1115.
[389] D.-h. Lu, W. L. Hase, J. Chem. Phys. 92 (1988) 3217.
[390] R. E. Wyatt, C. Iung, C. Leforestier, J. Chem. Phys. 97 (1992) 3477.
[391] S. Pedersen, J. L. Herek, A. H. Zewail, Science 266 (1994) 1359.
[392] E. Goldfield, Faraday Discuss. 110 (1998) 185.
[393] W. H. Miller, J. Phys. Chem. 99 (1995) 12387.
[394] W. H. Miller, Faraday Discuss., Chem. Soc. 102 (1995) 53.
[395] J. Qi, J. M. Bowman, J. Phys. Chem. 100 (1996) 15165.
[396] V. A. Mandelshtam, H. S. Taylor, W. H. Miller, J. Chem. Phys. 105 (1996) 496.
[397] T. C. German, W. H. Miller, J. Phys. Chem. A 101 (1997) 6358.
[398] K. Song, W. L. Hase, J. Phys. Chem. A 102 (1998) 3648.
[399] K. Song, W. L. Hase, S. Yu. Grebenshchikov, R. Schinke, J. Phys. Chem. A 106 (2002) 8339.
[400] C. Janssen, J. Guenther, K. Mauersberger, D. Krankowsky, Phys. Chem. Chem. Phys. 3 (2001) 4718.
[401] Y. Q. Gao, R. A. Marcus, J. Chem. Phys. 116 (2002) 137.
[402] M. B. Reyes, B. K. Carpenter, J. Am. Chem. Soc. 120 (1998) 1641.
[403] R. A. Zubarev, N. L. Kelleher, F. W. McLafferty, J. Am. Chem. Soc. 120 (1998) 3265.

Subject index

ISBN 0-444-50893-7

www.ingramcontent.com/pod-product-compliance
Ingram Content Group UK Ltd.
Pitfield, Milton Keynes, MK11 3LW, UK
UKHW050945290926
15470UKWH00005B/1653

9 780444 508935